HAWAI'I

Graphic Index of Maps and Views

1 INCH EQUALS APPROXIMATELY 21 MILES/34 KILOMETERS/1:1,300,000

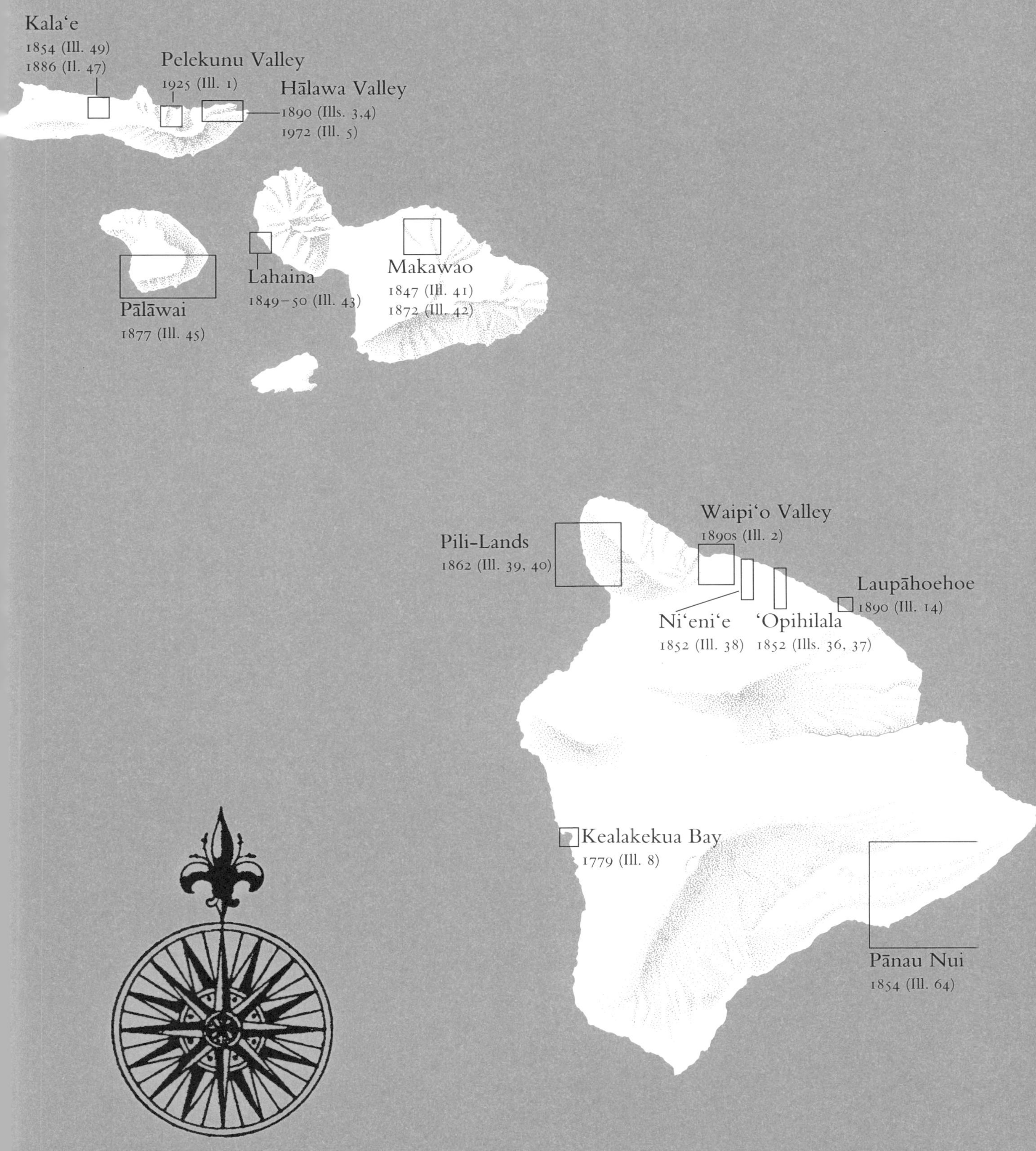

Surveying the *Mahele*

Oreehoua
ONEEHEOW
ATOOI
Wymoa Bay
WOAHOO

VOLUME 2

Palapala'āina

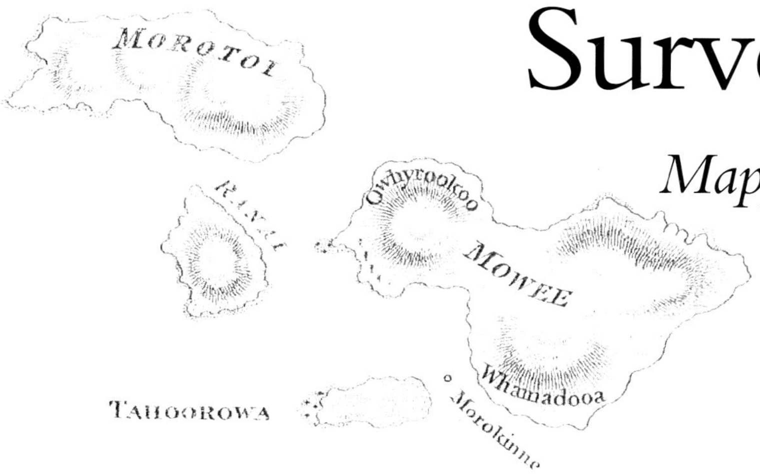

Surveying the *Mahele*

Mapping the Hawaiian Land Revolution

RILEY M. MOFFAT
AND GARY L. FITZPATRICK

An Editions Limited Book

Library of Congress Cataloging in Publication Data

Moffat, Riley Moore, 1947–
Surveying the *mahele* : mapping the Hawaiian land revolution / Riley M. Moffat and Gary L. Fitzpatrick.
Honolulu : Editions Limited, 1995.
120 p. 37 cm.
(Palapala'āina ; v. 2)
Includes bibliographical references and index.
ISBN 0-9150-1317-7
1. Surveying—Hawaii—History—19th century. I. Fitzpatrick, Gary L., 1947– II. Title. III. Series.
TA519.M64 1995 333.08
95-37752

∞ The paper in this book meets the guidelines for permanence and durability of the Committee on Production Guidelines for Book Longevitry of the Council on Library Resources.

Cover: *Honolulu, 1858,* Surveyed by J. W. Makalena, April 23, 1858. 30 x 38 cm.; Scale 1:4,572 (6 chains = 1 inch). (From Hawaii State Survey Division; document 1544, Drawer 15, no. 12. Ill. 52)
Title page: *Hawai'i,* Harrison & Co., 1784

Design by Archcetype Press
Washington D.C.

Published by Editions Limited
P.O. Box 10558
Honolulu, Hawaii 96816

Printed in Hong Kong

Contents

To my mother,

Who was fortunate enough to move to Hawai'i thirty years ago and introduce me to the wonderful world of the islands.

Riley

To Angela,

Who shares my appreciation for Hawai'i.

Gary

The authors wish to dedicate this volume to our mentor, Cline Mann.

Preface

IN THIS SECOND VOLUME of the series, *Palapala'āina*, we have focused on a particularly fascinating aspect of the history of the mapping of Hawai'i. The surveying and plat-making that were part of the process commonly known as the *mahele* was essentially the first cartographically related activity that was conducted because of and for the use of the people living in Hawai'i. The *mahele* itself was one of the most significant events in Hawaiian history; indeed, one could argue that the *mahele* was the most pivotal factor in altering the political, economic, and social landscapes of Hawai'i.

In this volume, our interests in surveying are several. In one respect, surveying is the most detailed form of mapmaking possible and is therefore the underlying framework of many forms of maps. Surveying is also a foundation of the western concept of private ownership of land, and we have examined how the concepts and practices of surveying affected the transition from the traditional Hawaiian land tenure system to the foreign concept of privately owned land. Lastly, the surveying work done during and in support of the *mahele* set the stage for—or created the need for, to be more accurate—the establishment of the Hawaiian Government Survey in 1870.

Numerous comments from readers of the first volume, *The Early Mapping of Hawai'i*, indicated that the surveying of the *mahele* is of great interest to many people. Indeed, we were somewhat surprised to find that the earliest mapping of Hawai'i was probably of secondary interest to the mapping of the *mahele* and the Hawaiian Government Survey. The recent activities associated with the centennial observance of the overthrow of the Hawaiian monarchy and the movement for some form of Hawaiian sovereignty underscored the significance of the *mahele*, and interest in this seminal event.

Because our expertise is the realm of current and historical cartography, we have attempted to avoid commenting on many important issues surrounding the *mahele* that do not relate to surveying and mapping. But surveying and cartography do not function in a vacuum; they are human activities and are therefore influenced by a variety of human factors. In recent years, cartographic historians have taken a more active interest in the cultural and social aspects of mapmaking. Debate has focused on such things as the role of mapmaking in the appropriation of the lands of the Native Americans, the relationship of cartography to military establishments and objectives, and mapmaking as a medium for the communication of culture, etc. Such studies have shown that cartography, as a form of communication, both reflects and influences the society in which it exists. In that light, we hope that our efforts provide new insights into the study of the *mahele* and encourage a broader appreciation of the survey plats and reports that document this profound transformation of land tenure in Hawai'i.

Terminology

Since this volume was written for two distinct audiences—those interested in Hawai'i and those interested in the history of cartography—it is important to clarify our usage of certain terms.

In general, Hawaiian words are shown in italics. Singular and plural forms are identical. Hawaiian place names still in current use and personal names are not italicized.

The word *mahele* is treated as any other Hawaiian term. We have opted to refer to all events generally associated with the change in the Hawaiian land tenure system by this word, but not to capitalize it or add a subjective adjective, as in "The Great Mahele," which is used frequently in histories of Hawai'i. We prefer to let readers assign its cultural and historical importance for themselves.

We have chosen to use "land" instead of the Hawaiian term *'āina* for two reasons. First, land is a word common to several of the European societies out of which surveying and private ownership of property evolved. A major use of the term land is defined in relation to ownership: "a portion of the earth's solid surface distinguishable by boundaries or ownership...." In terms of surveying, therefore, land is a more specific and appropriate term than the Hawaiian *'āina*. Secondly, *'āina* is a more complex word—both linguistically and culturally—than land. To speak of *'āina* is not to speak of just a certain piece of earth, for the word also invokes rich concepts of the relationship of people to the earth and the sustenance it provides.

Acknowledgments

ONE OF THE PLEASURES of writing a book is the making of new friends. To all those listed here, we want to say that we not only acknowledge your contributions, but we deeply appreciate the interest you have shown in our project and the courtesies you have extended to us. *Mahalo* for all that you have done.

Without slighting the contributions of others, it must be said that the knowledge of Cline Mann has been the most important resource available to us. In nearly a half-century as a practicing land surveyor in Hawai'i, Cline has developed an encyclopedic knowledge of land matters in the islands. In the research stage, he directed us to important documents, particularly surveys that could be found in the vault of the Hawai'i State Survey Office. His enthusiasm for the concept of this volume was most evident in his evaluation of our text, during which he challenged facts and ideas as well as encouraged us to expand on original lines of thought that we had developed. One of his most significant contributions was conveying to us a sense that the Hawaiian land tenure system was far more complex than the descriptions we encountered in the existing literature.

It is with pleasure that we acknowledge once again the contributions of Dr. Paul Markham Kahn. In countless ways he has encouraged us in general and assisted us in numerous particulars. In addition to sharing resources from his own library, which is now a part of the State Archives of Hawaii, he has alerted us to the existence of important source material. Finally, by reading and commenting on the text at various stages of development he significantly shaped the nature of the volume.

A book such as this cannot be done without the aid of the librarians and archivists who care for the cartographic and literary heritage of Hawai'i. At the Hawaiian Mission Children's Society, Stephanie Fitzpatrick, Marilyn Reppun, and Judy Kearney were most helpful in finding source material and arranging for the reproduction of illustrations. Ms. Reppun was also kind enough to arrange for us to visit a farm in which historic taro fields have been returned to production. Barbara Dunn of the Hawaiian Historical Society provided valuable assistance in finding and using material from that organization's library.

In map libraries, the most important resource is the staff, whose personal knowledge of the collections is more valuable than any card catalog. Thus, we are deeply indebted to Mabel Suzuki and Ross Togashi for making available to us the cartograpic resources of the Hamilton Library at the University of Hawai'i.

We are also thankful to other University of Hawai'i staff. Dr. Everett Wingert, Department of Geography, allowed us to use illustrations from his personal collection. We are also appreciative for the thoughtful comments about our manuscript. Dr. Chieko Tachihata, Curator of the Hawaiian Collection at the Hamilton Library, also took the time to read and provide very insightful comments on the final draft of this book. Puakea Nogelmeier, professor of the Hawaiian language, shared with us the fruits of his own research into matters relating to the historic lands of Hawai'i.

At the Bishop Museum, Betty Kam and DeSoto Brown were very helpful in navigating the rich photographic resources of that institution and for arranging to have reproductions made.

At the Hawai'i State Archives, Susan Shaner and Allen Hoof spent considerable time helping us find and reproduce documentation from the *mahele* era. Their intimate familiarity with textual sources and historic maps and surveys significantly aided our research efforts. We also thank Mary Ann Akao, the State Archivist, for granting permission to reproduce pages from the *Buke Mahele*, one of the great treasures of Hawaiian history.

We are also indebted to Paul Nuha and Kazutaka Saiki, the State Surveyors of Hawai'i, who allowed us to explore the contents of their agency's archives, which houses thousands of surveys dating back to the *mahele* era. The most important illustrations in this volume come from that agency. Although he has retired and moved away from the islands, the contributions of Charlie Okino cannot go unmentioned. In the early stages of our research, Mr. Okino provided valuable assistance in using the materials in the State Survey office.

Barnes Riznick and Robert Schleck of the Grove Farm Homestead Museum on Kaua'i took time from their duties to share with us the resources of that institution and to alert us to other sources of important information.

1 Pelekunu Valley, Moloka'i (circa 1925)

In this aerial photograph from the 1920s, evidence of habitation can be seen along the small coastal plain at the mouth of the valley. Several traditional land divisions are located here, including *apana*, or parts, of larger *ahupua'a* lying beyond the ridge line. This valley on the rugged north side of Moloka'i represents one of the many landscape types to which Hawaiians adapted their culture. The last permanent residents left in 1919.

Source: Hawai'i State Archives, negative number 21,727.

1 The Hawaiian Use of Land

THE *MAHELE* is rightfully considered one of the most significant chapters in the modern history of Hawai'i. Several legislative acts during the period 1845–1855 codified a sweeping transformation from the centuries-old Hawaiian traditions of royal land tenure to the western practice of private land ownership. In the first stage of this process, the Hawaiian monarch Kauikeaouli (1813–1854), also known as Kamehameha III, divided up his lands among the highest ranking *ali'i* (chiefs), *konohiki* (land managers), and favored *haole* (foreigners). Thus, the political and social relationships that had been the foundation of Hawaiian culture were permanently altered. Later, the *maka'āinana*, or common people, were permitted to acquire legal title to land they had tilled and lived on and to purchase other lands from the government. Finally, near the end of the *mahele*, one of the most significant changes occurred: granting naturalized foreigners the right to purchase land in Hawai'i.

It is not hyperbole to suggest that the impact of the *mahele* was far more profound than anyone, especially the Hawaiian people, ever dreamed. The *mahele* set the stage for a variety of alterations in the social, political, economic, and physical landscapes of Hawai'i. While the several events that comprised the *mahele* only involved changes in the pattern of land ownership, or tenure, historians have always viewed the *mahele* in a much broader perspective.

As will be explained in subsequent chapters, private land ownership necessitated the introduction of land surveying on a vast scale. In nearly thirteen thousand transactions, approximately two million acres were transferred to various *ali'i* and *maka'āinana*. The lands and parcels of property involved were described in a variety of ways: by name only, through textual descriptions, or graphically in the form of survey plats. It can be argued that surveying is a distinctly separate discipline from cartography, but the graphic portrayals of land compiled by surveyors have much in common with maps of all types. The surveys drawn during the *mahele* are deceptively simple, but they contain much information about the original tenure system used by the Hawaiian people.

It is impossible to evaluate the surveying work that occurred during and immediately after the *mahele* without an understanding of the conventions and customs that governed occupancy and use of both the land and waters of the islands. In no way, however, can a single, short essay such as this convey all the nuances of the Hawaiian concept of land.

Though much has been written about the land tenure system, recently published research suggests that much remains to be learned through archaeological investigation and reinterpretation of the oral tradition that is an important source of Hawaiian history.[1] It is apparent that the writings of earlier—and very reputable—scholars presented too simplistic a picture of Hawaiian land use. Revelations from field archaeology, careful review of historic documents, intensive study of *mahele* records, and even efforts to revive the ancient *lo'i*—or cultivated taro patches—all point to a very complex and dynamic system.

The magnitude of the decline of the Hawaiian population after the arrival of Europeans is well-documented. Unfortunately, much of the knowledge about ancient Hawai'i perished with the people. Working from even the more conservative population estimates made at the time of Captain James Cook, it is apparent that the Hawaiian

2 *Waipi'o Valley, Hawai'i (circa 1890s)*

Once a major center of Hawaiian population, the large valley that marks the boundary of the Kohala and Hāmākua districts on the island of Hawai'i had become sparsely populated just a few decades after the *mahele*. Well-watered and relatively flat, Waipi'o offered ideal conditions for the cultivation of taro, the staple of the Hawaiian diet.

Source: Bernice Pauahi Bishop Museum, Honolulu, Hawai'i; negative number CB 23530.

people had been reduced by at least three-quarters by 1853, near the end of the *mahele*. It is impossible to say how much information about land tenure disappeared. Since there is significant variation in landscapes throughout the islands, however, it is reasonable to assume that there were many localized practices that were never recorded and thus forgotten after the well-documented decline of the Hawaiian population. Descriptions of the pre-*mahele* land tenure system that imply its universal application throughout the islands may not in fact represent a true picture of Hawaiians' relationship to the land.

The following overview paints a broad picture of land use and governance in Hawai'i. Previously published works describing Hawaiian land tenure are heavily relied upon, but one particular facet of traditional scholarly writing about Hawaiian society proved bothersome *vis à vis* surveying. A common feature of many writings, especially the general histories of the islands, is the reliance on terms from European feudalism to explain the social structure of Hawaiian society. Many texts utilize such terms as king, nobility, priesthood, and serfs.

The feudal analogy would not be significant except for the fact that the land system that supplanted the original Hawaiian land practices grew directly out of the dissolution of medieval Europe. Using terms derived from the origin of Western land philosophy to describe the Hawaiian system misleadingly implies that component parts and processes were identical.

The Physical Environment and Land Use in Hawai'i

Hawaiians depended solely on the local environment for their food and utilitarian products. There was no importation of goods to supplement what they obtained from the islands and adjacent waters. Land, therefore, was the very foundation of existence. Without access to land—

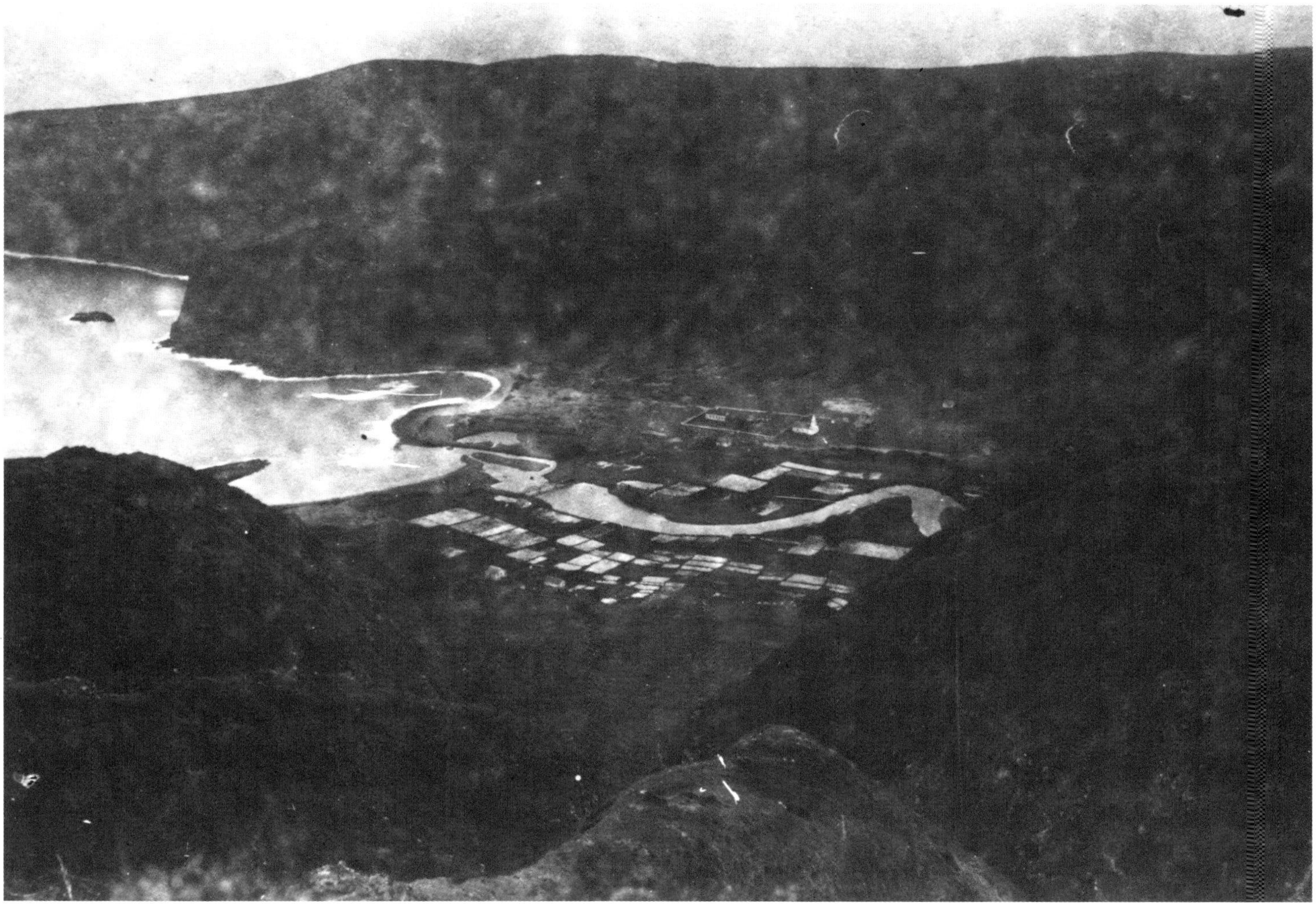

3 Hālawa Valley, Moloka'i (circa 1890s)

This photograph of the mouth of Hālawa Valley shows how irrigated fields were laid out near reliable sources of fresh water. Unfortunately, even photographs as early as this may not reflect the way Hawaiians first used the land, for the introduction of rice farming and other Western practices could have altered the landscape by this date.

Source: Bernice Pauahi Bishop Museum, Honolulu, Hawai'i; negative number CP 104,294.

and to the waters considered an integral part of the land—there was no way to sustain life.

Hawaiians took full advantage of the natural resources around them, utilizing flora, fauna, and minerals from the seashore to the peaks. Quarries at the summit of Mauna Kea, for instance, provided obsidian for knives and other tools. Upland forests yielded trees for dugout canoes, timber for other uses, birds for ornamental featherwork, and plants for food and medicinal purposes. Lower, grassy areas provided thatching for houses. Stones for platforms and walls could be found virtually anywhere. The Hawaiians fished the waters off the islands and built artificial fishponds in shallow reef areas. They displayed great ingenuity in adapting the available natural resources to their needs. The environment did not dictate the nature of Hawaiian society so much as it presented options and limits. Many aspects of Hawaiian society, therefore, can be seen as adaptations to the physical characteristics of the islands.

Isolated at the northern edge of the Pacific tropics, the Hawaiian Islands enjoy climatic conditions ameliorated by temperate ocean waters and the moderate trade winds that predominate for most of the year. Due to the relatively high altitude of the major inhabited islands, clouds moving over the peaks and ridges deposit rain on the windward sides. Climatic disruptions to the pattern only occasionally bring rain to the leeward sides.

Each island has a unique appearance, primarily due to its age. The older islands have been more heavily eroded and therefore exhibit landforms that are not a direct result of volcanic activity. While each island has a range of wet to dry landscapes, the major features—peaks, slopes, valleys, plains, and reefs—vary in size, shape, and distribution from one island to the next.

The biological potential of the islands is as varied as the physical environment. The combination of wet-dry and low-high ranges of environment offered a wide variety of habitats for flora and fauna that reached the islands through both natural dispersion and human introduction. Prior to

4 Hālawa Valley, Moloka'i (circa 1890s)

In this image looking toward the twin waterfalls at the head of the valley, it can be seen that just a century ago such broad valleys did not exhibit the lush vegetation that is often found today. The introduction of foreign plants and the abandonment of traditional agricultural practices due to population decline and migration significantly altered the appearance of much of Hawai'i. During the *mahele*, many *maka'āinana* claimed land that they had farmed in fertile valleys such as this.

Source: Bernice Pauahi Bishop Museum, Honolulu, Hawai'i; negative number CD 22719; J. F. Rock.

5 *Hālawa Valley, Moloka'i, 1972*

Remains of terraced fields can still be found beneath the dense canopy of Hālawa Valley. The extensive agricultural works suggest that even without the benefit of private ownership, Hawaiians invested tremendous effort in developing and maintaining property to make it more productive, contradicting the opinion of many Westerners that there was no incentive for Hawaiians to improve their land under the native land tenure system.

Source: Gary L. Fitzpatrick.

the arrival of humans, biological diversity was limited by the isolation of the islands, over two thousand miles from the nearest continent and nearly as far from the closest archipelago. Scholars disagree about what the vegetation pattern was like when Hawaiians first arrived in the islands. But there is little doubt that much of the lush vegetation that characterizes Hawai'i today was not present when the Europeans landed. In fact, early photographs show the islands to be far more barren than they are today. Humans also altered the environment through the introduction of useful plants and animals in their migrations to the islands, which began at least by 600 A.D., if not several centuries before. Europeans arrived in 1778, bringing a wide array of species, including ornamentals and pests as well as useful plants.

This is not to imply, however, that Hawaiians simply "lived off the land," for they managed and altered their environment to improve the quality of their lives. Agriculture and fishing were the mainstays of the economy. Although some scholars argue that early Hawaiians practiced an environmentally destructive form of slash and burn farming,[2] they ultimately developed an exceptionally complex and efficient system in which taro was the staple crop. Even today, underneath the vegetation that has overgrown so many areas, the remains of the Hawaiian agricultural system are impressive in their extent and design. In addition to cleared land for crops, there is evidence of widescale construction, including houses, enclosures of rock and wood used for various purposes, large stone platforms for religious ceremonies, terraces and irrigation systems, and an extensive complex of artificial fishponds. In short, there is considerable evidence that Hawaiians changed the physical landscape to suit their needs.

6 Kalokohanahou fishpond, O'ahu, 1930

Fishponds, such as this one adjacent to Pu'upahu in the Kāne'ohe area, were an integral part of the traditional Hawaiian landscape and are still readily evident, especially on Moloka'i and O'ahu. The rights to use fishponds and certain fishing grounds were well-regulated in traditional Hawaiian society though no one owned them.

Source: Bernice Pauahi Bishop Museum, Honolulu, Hawai'i; negative number CN 15361; J. Gilbert McCallister.

Because Hawaiians were such accomplished cultivators, population distribution in the islands was influenced but not completely limited by rainfall. Streams originating in the mountains brought water through drier coastal areas, making it possible to practice wetland agriculture via irrigation systems in areas that did not receive much direct rain, such as the plains of Waikīkī or valleys like Anahulu and Mākaha on the leeward side of the Ko'olau Mountains of O'ahu. Archaeological evidence, oral tradition, and written commentaries indicate that the bulk of the population lived along the coast. Much of the non-coastal population settled in the many fertile valleys found throughout the islands.

For this volume, the key issue is where people were located at the time of the *mahele*. As will be fully explained later, *maka'āinana* could only claim land they had been using at least two years. Consequently, most of the small parcels of real estate that passed into private hands, known as *kuleana* lands, were prime agricultural land. A fairly reliable picture exists of the distribution of people at the time of the *mahele*. In 1931, a University of Hawai'i geography professor named John Wesley Coulter produced a series of maps reflecting population distribution on each of the major islands at the end of the *mahele* era. His maps were based on the comprehensive and uniform census of 1853, in which population counts were recorded by administrative units smaller than a whole island.

Coulter's maps of population distribution have been reproduced here because they mark the mid-point of the relatively short time period examined in this volume. He compiled three maps for each island, it being his intention "to show the relation between the density and distribution of people and their use of the land and sea, uses which were a direct reflection of the natural environment of the islands."[3] As examples of his approach, the rainfall and relief maps of Maui also have been reproduced. The three Maui maps indicate that population was not a simple reflection of rainfall. As their distribution shows, Hawaiians farmed dry as well as wet areas, and they made extensive use of the resources of the sea. Most of the seventy-one thousand

Hawaiians in 1853 lived in areas of moderate rainfall—it simply was not pleasant or healthy to live permanently in the wettest regions. A significant number of people were also found in some areas of low rainfall, where they had access to streams, springs, or wells.

Broadly speaking, Coulter's maps represent the demographic landscape that the surveyors of the *mahele* era encountered. It was not truly indicative of the original uses of land by Hawaiians because of the extensive changes wrought during the previous seventy-five years of contact with Western culture, foreign plants and animals, and diseases. Nonetheless, the original Hawaiian pattern of land occupation formed the foundation of land tenure at the onset of the *mahele*. When the surveyors of the *mahele* took to the field, they were coping with island landscapes that had undergone significant alteration since the arrival of the first Hawaiians. Much of their work, therefore, involved interpreting the human use of the land within the context of numerous physical and biological factors. As will be seen later, many of the surveyors failed to appreciate the sophistication and complexity of the ways in which Hawaiians functioned within their environment.

The Social Organization of the Land

A general summary of the pre-*mahele* Hawaiian land tenure practices distilled from many published works about Hawaiian history might read something like this: All land belonged to the king; he kept the parcels he wanted for his own use and distributed major units of land to the higher chiefs; these chiefs in turn distributed the use of smaller parcels of land to lesser chiefs; at the bottom of the hierarchy, the commoner was granted the use of a particular plot of land to farm; land belonged only to the king and the use of it could be revoked at will.

Of course it was not that simple. But viewed broadly, such a description bears a superficial similarity to European feudalism, in which lords granted fiefs to vassals in return for pledges of loyalty and service. The differences, however, are quite profound, and they can be found at several levels of the social and economic hierarchy of Hawai'i.

The analogy to feudalism hinges partly on the use of the term "king," an imported English word. In old Hawai'i, the ruler was known as the *ali'i'ai moku*, which Kuykendall describes as being "*the* chief" or "the *great* chief."[4] In Pukui and Elbert's *Hawaiian Dictionary*, first published in 1957, *ali'i'ai moku* is defined as the chief of a *moku*, an island or one of the large traditional districts of an island.[5] Kirch and Sahlins cite that and a variation, *ali'i nui 'ai moku*, and claim that both variations represent a conquering chief and are descriptive of the position held by the chiefs who ascended to power prior to and after the arrival of Europeans.[6] They note that the term means "the great chief who ate the island or district."

The Hawaiian word most analogous to "king" is *mō'ī*, which Pukui and Elbert define as "king, sovereign, monarch, majesty, ruler, queen."[7] The word *mō'ī*, however, is apparently of comparatively recent origin, and does not appear to have been used by Hawaiians before the arrival of Europeans. In that sense, *mō'ī* can be seen as an appropriate title for the descendants of Kamehameha, but it is probably not the right term for describing the highest authority of the social structure in Hawai'i before the Europeans. If *ali'i'ai moku* refers to a chief who has assumed power either through warfare or some other form of usurpation, then there appears to be no word that describes the highest form of chief who assumed the position through some legitimate and peaceful means. In the absence of such a word, therefore, *ali'i'ai moku* is accepted here as the most appropriate for either means of ascension.

Without going to great lengths to explain the complexities and nuances of Hawaiian concepts of lineage, suffice it to say that primogeniture was not the rule of the land. Neither the title nor the lands of an *ali'i'ai moku* necessarily went to his eldest son. Indeed, the *ali'i'ai moku* could be female. In the decades prior to the arrival of Europeans, warfare apparently accounted for many of the changes in leadership. Whether the title was acquired peacefully or through force, accession to the position of *ali'i'ai moku* entailed rights and responsibilities regarding the land but did not confer absolute ownership. In this respect, Hawaiian society differed significantly from European feudalism, where the rights of land inheritance became absolute and were codified into law.

The differences between the Hawaiian and Western systems of land tenure can be traced to very basic religious beliefs. The argument has been made that in Hawai'i, land belonged to the gods, and that the *ali'i'ai moku* was merely the steward of the land.[8] In previous literature, therefore, the descriptions of the Hawaiian land tenure system err in stating that land "belonged" to the king. The Hawaiian concept of stewardship is quite different from Western practices, where "property rights" make land a commodity that individuals own and can do with as they please.

One of the key acts of a new *ali'i'ai moku* was known as the *kālai'āina*, or redistribution of lands. The control of large units of land was given to higher ranking *ali'i*, who in turn could redistribute sub-rights to lesser *ali'i*. In existing literature, it is not clear how people farther down the social ladder were affected by the *kālai'āina*. In any case, the *kalai'āina* essentially legitimized the rule of the *ali'i'ai moku* in the eyes of his followers. The custom also clearly emphasized that land was not owned outright by anyone.

In return for grants of land, the *ali'i* were obligated to provide services, food, and other products to the *ali'i'ai moku*. The matter of service to the *ali'i'ai moku* also invokes images of the king-lord, lord-vassal relationships that

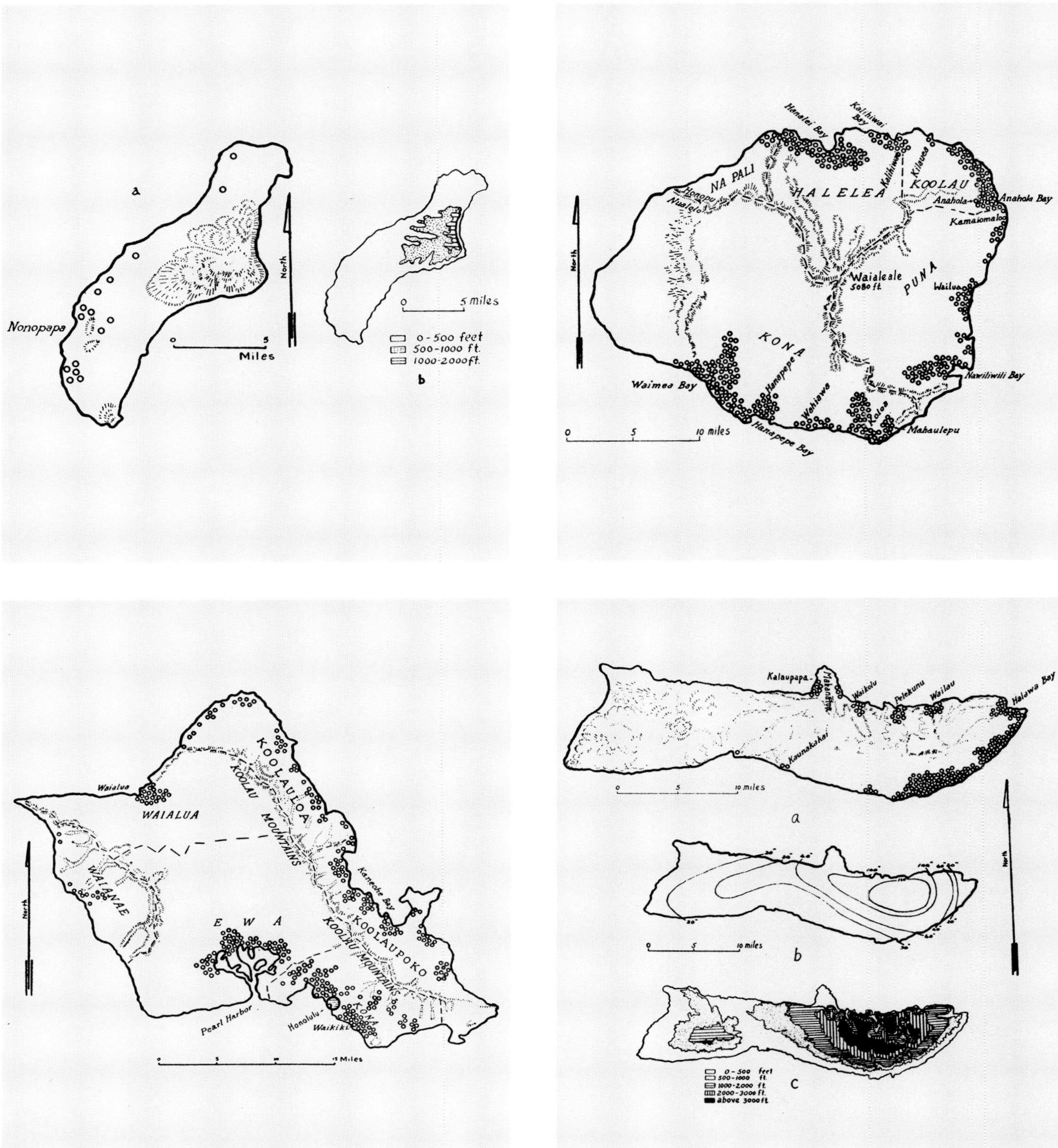

7 *Hawaiian population distribution in 1853*

These maps by the geographer John Wesley Coulter indicate the approximate distribution of population in the islands at the time of the *mahele*. For each island, Coulter prepared maps showing population distribution, elevation, and rainfall to illustrate the inter-relationship of these factors. All three maps of Maui are illustrated; for the other islands only the map of population distribution is shown. The small circles represent fifty people on the islands of Ni'ihau, Kaua'i, O'ahu, Maui, and Hawai'i, twenty people on Moloka'i, and ten people on Lana'i. Three towns are shown by larger circles: Honolulu, with 5,500 people; Lāhainā, with 2,750; and Hilo, with 500.

Coulter, John Wesley. *Population and Utilization of Land and Sea in Hawaii, 1853*. Honolulu: Bishop Museum, 1931.

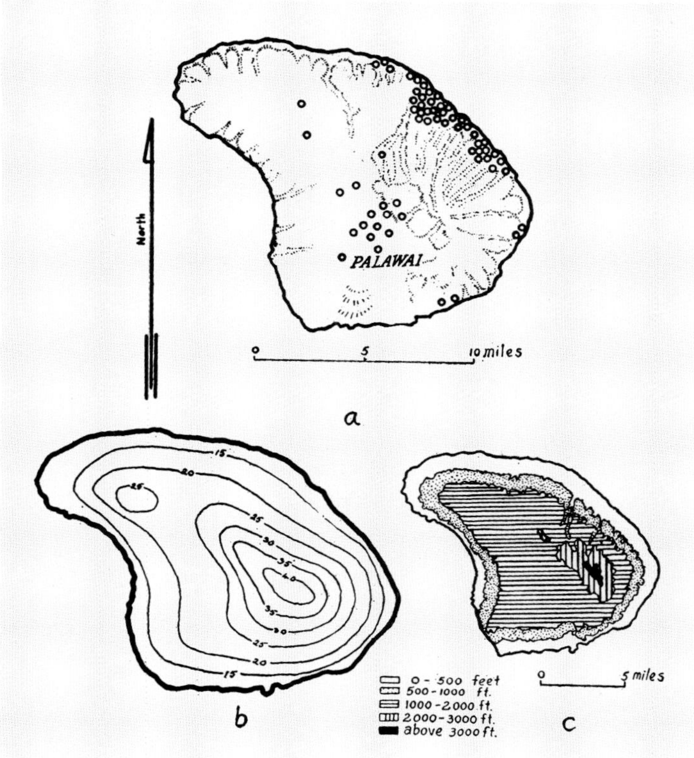
North
PALAWAI
0 5 10 miles
a
0 - 500 feet
500 - 1000 ft.
1000 - 2000 ft.
2000 - 3000 ft.
above 3000 ft.
0 5 miles
b
c

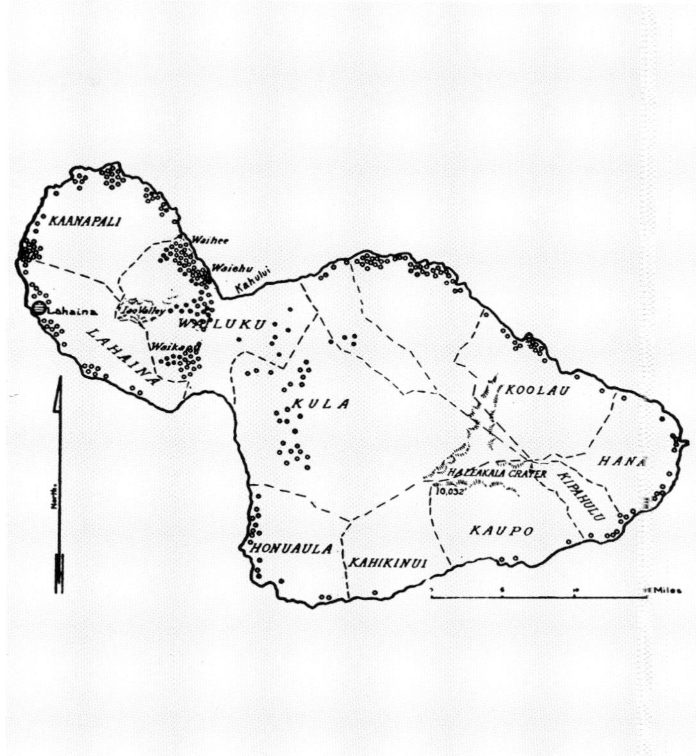
KAANAPALI
Waihee
Waiehu
Kahului
Lahaina
WAILUKU
LAHAINA
Waikapu
KULA
KOOLAU
HANA
HALEAKALA CRATER
10,032
KIPAHULU
KAUPO
HONUAULA
KAHIKINUI
Miles

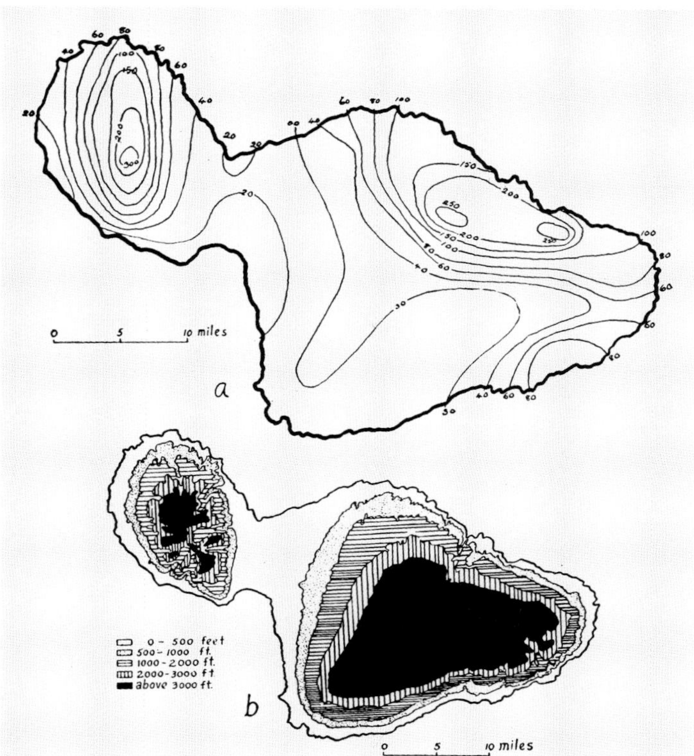
0 5 10 miles
a
0 - 500 feet
500 - 1000 ft.
1000 - 2000 ft.
2000 - 3000 ft.
above 3000 ft.
b
0 5 10 miles

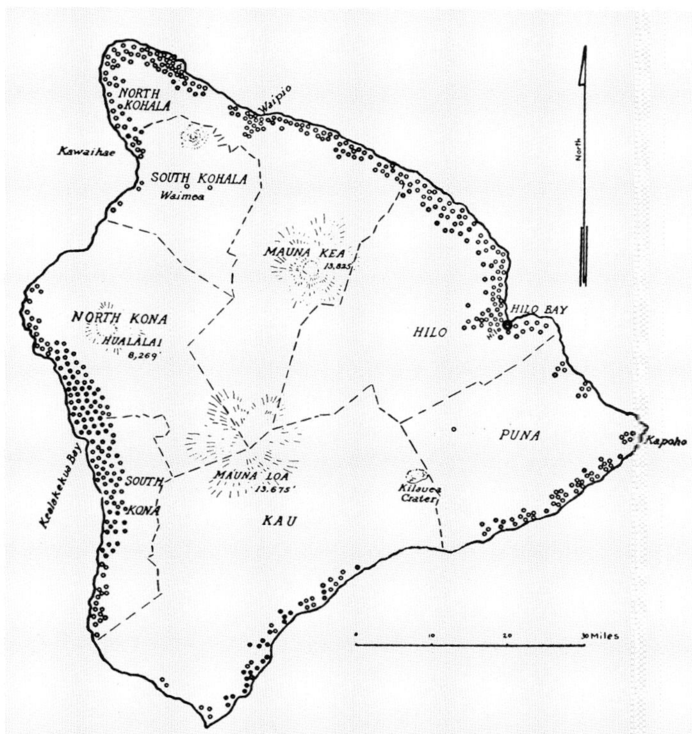
NORTH KOHALA
Waipio
Kawaihae
SOUTH KOHALA
Waimea
MAUNA KEA
13,825'
NORTH KONA
HUALALAI
8,269'
HILO
HILO BAY
PUNA
Kapoho
Kealakekua Bay
SOUTH KONA
MAUNA LOA
13,675'
Kilauea Crater
KAU
0 10 20 30 Miles
North

were a prominent part of European feudalism. The *ali'i*, the upper class of Hawai'i, performed most of the same functions as the European nobility; they were the warriors, the administrators, and the counselors. In certain respects, however, the *ali'i* did not have as many rights or advantages as their European counterparts when it came to land. Since they used the land only under the grace of the *ali'i 'ai moku*, who did not actually own it himself, they could not pass on land rights to their descendants. As a class, the *ali'i* had little recourse to legal justice. The word of the *ali'i 'ai moku*, was final.

Despite having no legal title to their lands, the *ali'i* of Hawai'i possessed power and lived well from the labor that others invested in the soil. But according to Western economic theory—especially as taught to the Hawaiians by the American missionaries—there was no incentive for anyone, from the *ali'i* on down, to invest significant effort in land that might not stay in the family. This contention will be examined more carefully in the next chapter.

The tasks of working the land and fishing fell to the *maka'āinana*, the commoners. Like the *ali'i 'ai moku* and the *ali'i*, the *maka'āinana* did not own land. Instead, the *ali'i* granted them plots of farmland and housing sites for their own use and access to other needed resources. In a society where life depended upon access to the land, having no property that was legally their own meant the *maka'āinana* were at the mercy of their landlords. They had to accept whatever terms were offered for the use of land or move on. But unlike vassals in feudal Europe, the *maka'āinana* had the freedom to move, to seek out better terms or conditions under a different *ali'i*.

According to some accounts, those terms could be rather harsh by modern standards. A large percentage of their produce was siphoned off—call it taxes, tribute, or rent—and they were expected to work on the land of their immediate landlord as well as that which belonged to the *ali'i 'ai moku*. Some scholars estimate that the people who worked the soil could keep only about one-third of what they produced, the rest going to support the *ali'i* and the *ali'i 'ai moku*.

In reality, however, life for the common people may have differed significantly from the way the system was supposed to work. Archaeological evidence suggests that at some point in Hawaiian history, the *maka'āinana* did not move frequently, and that families may have stayed in one area and worked the same land from one generation to the next.[9] Unfortunately, there is too little information dating from before the arrival of Europeans to make unequivocal statements about many aspects of the Hawaiian land tenure system.

Sanford B. Dole, one of the leading figures in the overthrow of the Hawaiian monarchy and the annexation of Hawai'i to the United States, stated in an 1892 paper that beginning in about the thirteenth century, Hawaiian society was drastically altered by numerous internecine wars.[10] Before then, he argued, life was relatively settled, and people actually may have possessed a concept of private land ownership. His thesis, however, suffers from the same lack of documentation that other scholars have faced. His contention, however, receives general support in the recent research of Kirch and Sahlins. Their archaeological work in Anahulu Valley in the Waialua district on O'ahu indicates that before about the twelfth century, people apparently lived in an area for at least several generations, although that may not necessarily mean that private land ownership was the basis of their society.

Conventional scholarship suggests that a basic cause of the disintegration of European feudalism was the development of a monetary society. Likewise, the introduction of currency and related attributes of the Western economy eroded the foundations of the Hawaiian land tenure system. Before there was a money economy, most *maka'āinana* had to work the land. Some were employed in special occupations—canoe maker, house builder, fisherman, etc.—but received in return only the necessities of life. On tropical islands lacking any source of metals, there were few material goods to accumulate. The common people worked for subsistence. They passed on very little to their children besides canoes, tools, utensils, a few artifacts, knowledge, and status in life.

In the post-European era, the introduction of a monetary economy paralleled the arrival of new economic opportunities. Since the *maka'āinana* were free to move, some chose to migrate to the developing port towns, where a variety of jobs were being created. A number of young men also went to work on the merchant and whaling vessels.

Money put the products of the land in a different light. Food once grown for sustenance was now sold for cash, by the *ali'i* and probably also by the *maka'āinana*. When Europeans learned that the island forests included sandalwood trees, a booming business began that brought substantial amounts of money to the *ali'i*. But the sandalwood trade brought problems, too: Harvesting the trees meant that less labor was available for growing foodstuffs; and the concept of long-term debt was introduced to the islands.

In the *mahele*, land was distributed to *ali'i* and *maka'āinana* as if Hawai'i was the traditional agricultural society it always had been. For the most part, land was divided by the traditional units ranging from districts to individual garden plots. But the surveyors who went into the countryside to make plats of these traditional land units were confronted by a different situation: People were moving away from the land, they were dying faster than they were being replaced, knowledge about the original land tenure system was disappearing, the *ali'i*—the principal landlords—were viewing land in a different economic sense, and the relationships between those who controlled land and those who worked it were becoming less harmonious.

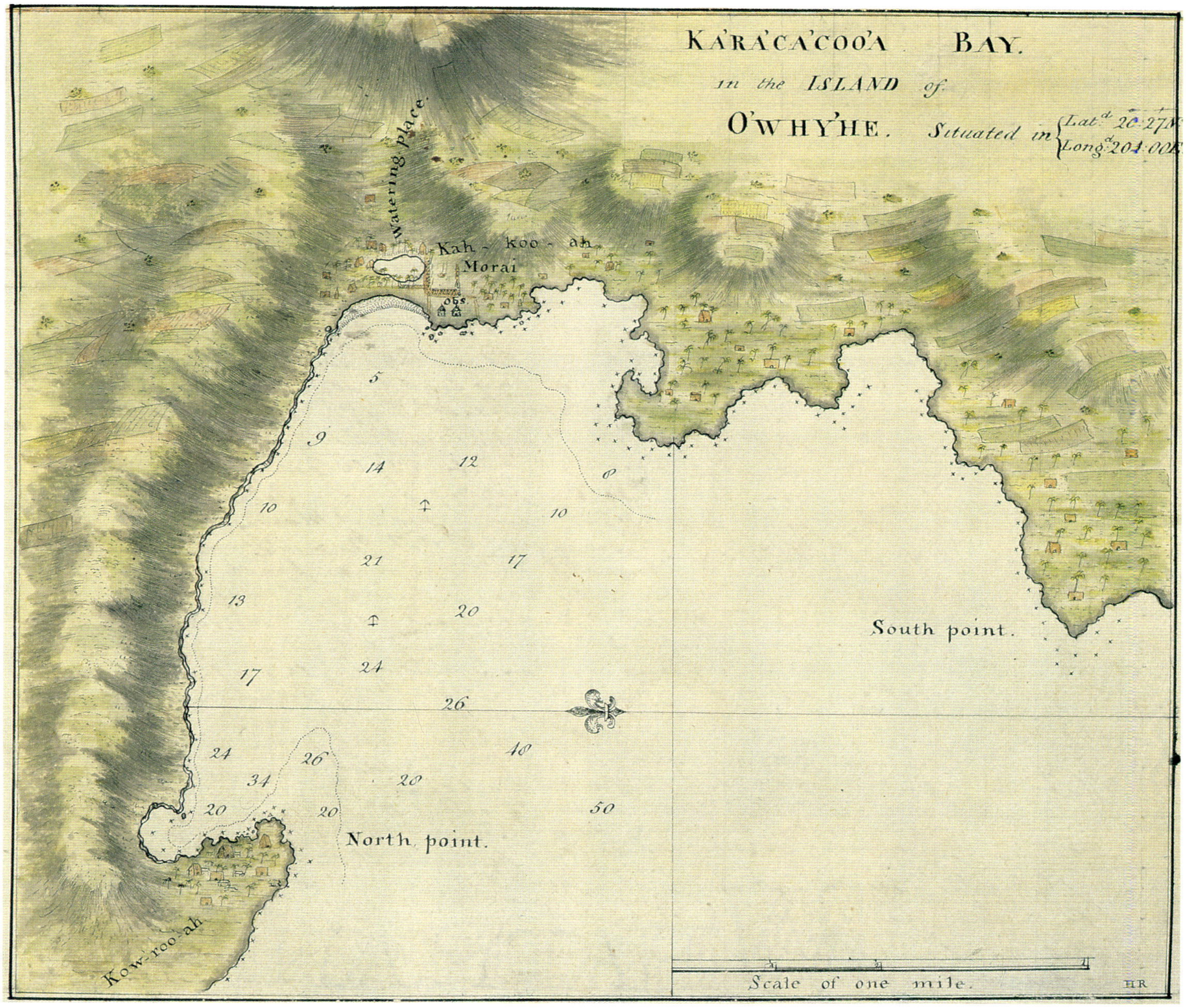

8 Kealakekua Bay, Hawai'i, 1779

This map by Lt. Henry Roberts from Captain Cook's voyage is among the earliest graphic portrayals of the field and village patterns of the Hawaiian system of land tenure. While such images are important documentary sources, a fair degree of generalization was probably employed by most cartographers of the era, so these maps cannot be taken as literal interpretations of land use by Hawaiians.

Roberts, Henry. *Ka'ra'ca'coo'a Bay, in the Island of O'why'hee.* ms. map, 18.6 x 21.1 cm. Scale ca. 1:24,700. In No. 2 Log of Henry Roberts Mate; DL ref. MSQ 152, page 45. Source: from the original manuscript chart in the Dixson Library, State Library of New South Wales.

Moku *and* Kālana

To say that all land was controlled by the *ali'i'ai moku* is somewhat misleading, because until the time of Kamehameha there may have been only a few short periods when all the islands were under the control of one person. For much of Hawaiian history, several *ali'i'ai moku* controlled large units of land concurrently. At least for some time prior to Kamehameha, warfare was rather frequent, and control of major portions of islands shifted with the tide of battle. Until Kamehameha brought all the islands under his control in the 1790s and early 1800s, an *ali'i'ai moku* might control a major portion of an island, one or more whole islands, or some combination thereof.

For centuries, Hawaiians have identified by name those districts that *ali'i'ai moku* might have controlled. Several of these names were among the few cultural features indicated on the early maps of Hawai'i, such as that by Lieutenant Henry Roberts of Captain Cook's third voyage (*EMH* Ill. 2) and Lieutenant Joseph Baker on Vancouver's voyage (*EMH* Ill. 21). For the island of Hawai'i, foreign-

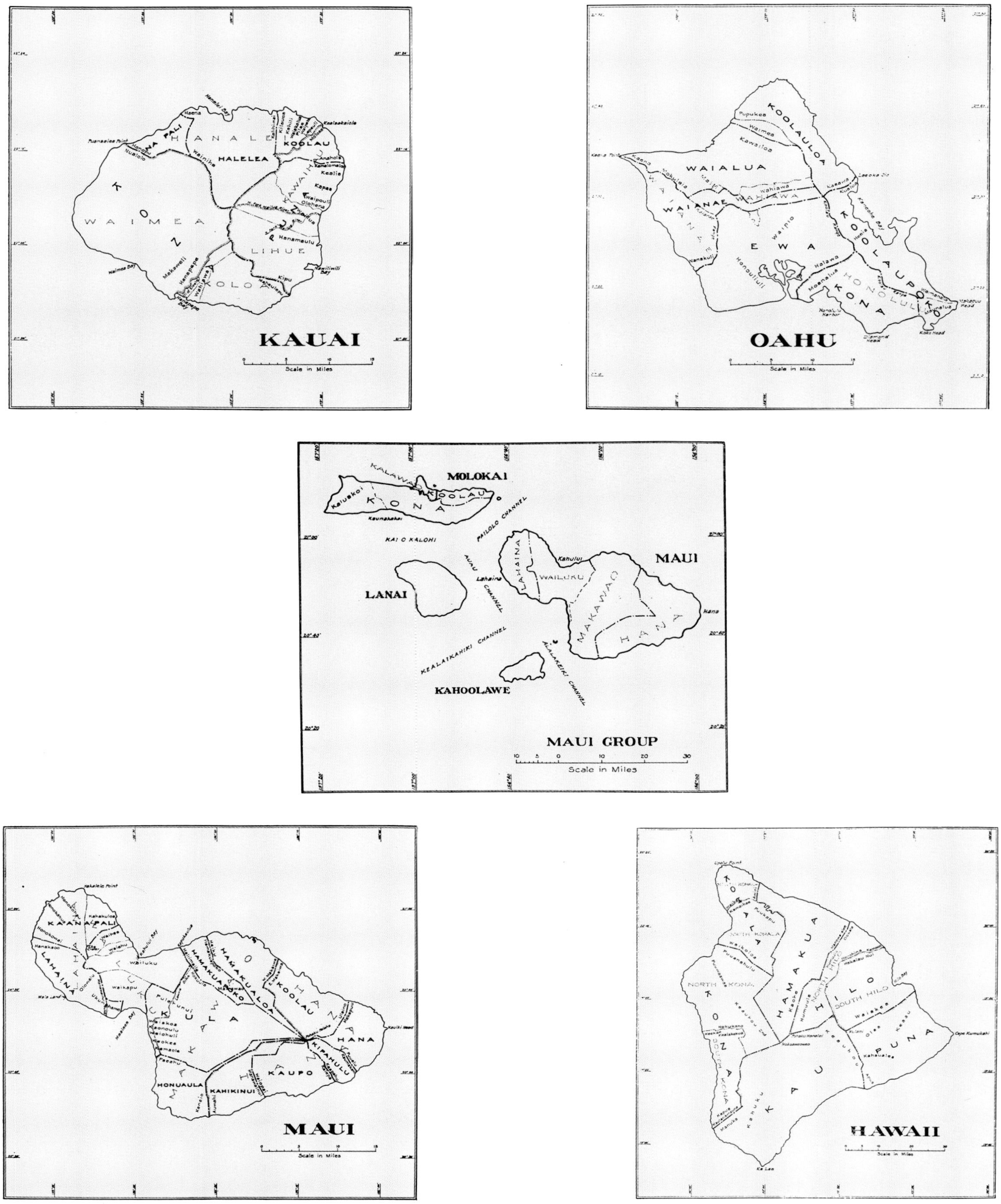

9 *Administrative maps of the islands of Hawai'i, 1935*

John Wesley Coulter produced these maps to illustrate both modern and historic *moku*, or districts, in Hawai'i. *Moku*, also known as *kalana*, were the major administrative units in the traditional Hawaiian land tenure system.

Coulter, John Wesley. *A Gazetteer of the Territory of Hawaii*. Honolulu: University of Hawaii, 1935.

ers managed to record all of the district names, As noted in *The Early Mapping of Hawai'i*, the spellings on these maps are transcriptions of the sounds heard by European navigators. Since the Hawaiian language was not reduced to writing until the mid-1820s, there was no standardization of spelling on the early maps.

An interesting aspect of the political geography of the Hawaiian land tenure system is that the language provides little distinction between an island as a whole and a district or region. The Hawaiian word *moku* has multiple meanings, and to some extent it is unclear whether current usage is consistent with historical use of the term. In the modern context, *moku* refers in one sense to a district or region of an island. In another sense, it is the word for island. Pukui and Elbert's dictionary defines the word as, among other things, "district, island, islet, section. . . ."

There is another word, *mokupuni*, to which the two lexicographers also ascribe the meaning "island." Lorrin Andrews, who published the first authoritative dictionary of the Hawaiian language in 1846, noted that *mokupuni* means a division of land surrounded by water. According to Samuel Kamakau, a noted Hawaiian scholar educated by Andrews at the missionary school at Lahainaluna, Maui, the word for island was *'āina* and that for a district was *moku'āina*.[11] Kamakau also notes two related terms, *kālana* and *'okana*. He implies that *kālana* is equivalent to *moku'āina* and that *'okana* is a subdistrict.

In historic usage, however, another meaning of *moku* should be cited. In a letter written in 1858, Lorenzo Lyons noted that

> . . . the Moku in ancient times was the Commons—the people had a perfect right to anything on it—without taxation.[12]

The common land to which Lyon referred was situated above most of the *ahupua'a* and encompassed forest and barren mountain. Although the land may have technically belonged within a given *ahupua'a*, it was apparently available to all the people living within that district. In that sense, *moku* may have come to refer to the portion of an island that shared access to a single common area.

Given this historical meaning, *kālana* may have been the correct term for what is called *moku* today. It is worth noting that there is frequent reference to the term *kālana* in historical documents, including those by the noted surveryor and government official Samuel P. Kalama, about whom much is said in later chapters. Accepting only the narrower meaning of *moku* as a commons, however, calls into question the meaning of *ali'i'ai moku*. Rather than debate the relative merits of the various terms that can be applied, this volume will use the term *moku* simply to maintain consistency with other writings that deal with land issues. Nonetheless, the discussion points out that there is still some uncertainity in the knowledge about ancient land divisions in Hawai'i.

Depending on one's choice of situation and terminology, an *ali'i'ai moku* could therefore have controlled a *moku*, *mokupuni*, *'āina*, *moku'āina*, or *kālana*.

In the context of surveying and land divisions, *moku* is usually taken in the sense of being a major portion of an island that is delineated by significant topographic features or climatic zones. About six hundred years ago, according to the dating of surveyor Curtis J. Lyons, the Hawaiians created the *moku* and settled on a series of names for them.[13] Kamakau wrote that the divisions were made at that point in the past when an island became heavily populated.[14] The original *moku* have survived, although with some modifications, as administrative districts in modern Hawai'i. In 1935, Coulter published a series of maps showing the relationship of the original *moku* to their modern counterparts, four of which are reproduced here.

Three of the major islands were divided into six *moku* each. Hawai'i had the districts of Puna, Hilo, Hāmākua, Kohala, Kona, and Ka'ū. The *moku* of O'ahu were Kona, Ko'olau Poko, Ko'olau Loa, Waialua, Wai'anae, and 'Ewa. The six *moku* of Kaua'i were Kona, Puna, Ko'olau, Halele'a, Nāpali, and Waimea.

According to Coulter, the other islands are not as straightforward. Ni'ihau was considered a separate *mokupuni* but had no internal districts or *moku*. The central islands of the group are rather complex. Lana'i was sometimes considered a *mokupuni*, but in some references it was considered a *kālana* subordinate to one of the land divisions on the island of Maui. Kaho'olawe is generally considered a *mokupuni*, but at least one reference places it as a *kālana* subordinate to the *moku* of Ko'olau in Maui. Traditionally, Moloka'i was divided into two *moku*, Ko'olau on the north side and Kona on the south. Some portions of Moloka'i have been described as being in a *kālana* called Kālana, and most of the western part of the island was said to be in the *kālana* of Kaluako'i.

Maui is clearly the most complex of the islands in terms of *moku*. In the most common description, Maui consists of twelve *moku*. Traditionally, the western portion of Maui was divided into two *moku*, Lahaina and Kā'anapali. Other clearly defined *moku* were Hāmākua Poko, Hāmākua Loa, Ko'olau, Hāna, Kīpahulu, Kaupō, Kahikinui, Honua'ula, and Kula. Two areas, known as *ahupua'a*—which will be discussed in the next section—lay within the boundaries of Lahaina but were considered separate *kālana*. Two other *ahupua'a* named Wailuku and Waikapū were considered to be in a smaller district called Na Poko, although some references have these comprising the twelfth *moku* of Wailuku.[15]

Confusion about the status of the various *moku* on Maui

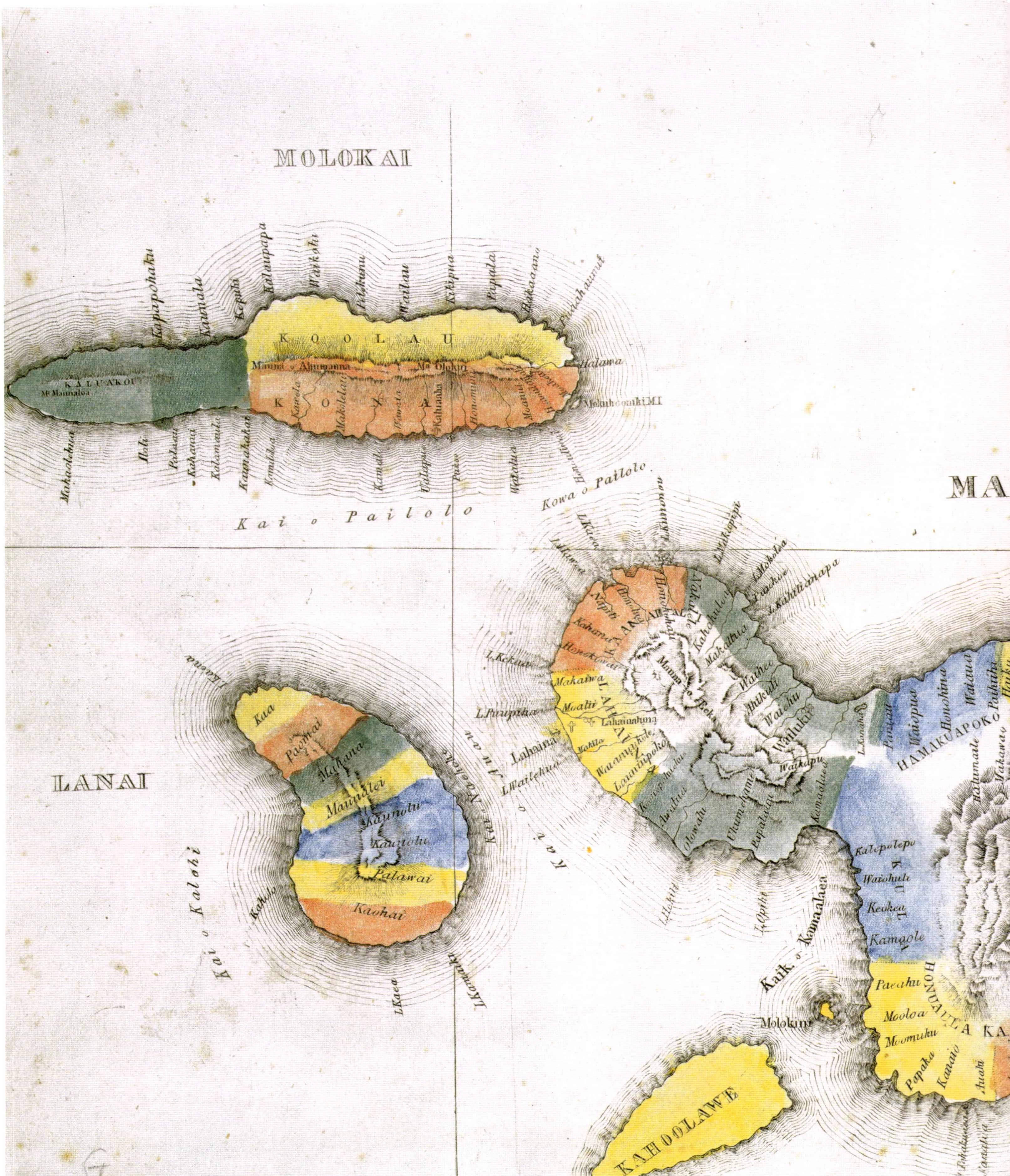
MOLOKAI
LANAI
KAHOOLAWE
MA
KOOLAU
KONA
KALUAKOI
Kai o Pailolo
Kowa o Pailolo
Kai o Kalohi
Molokini
Halawa
Lahaina
Kaunolu
Palawai
Kaohai
Kamaole
Kalepolepo
Waiohuli
Keokea
Paeahu
Mooloa
Moomuku
Papaka
Kanaio
Wailuku
Waikapu
Waihee
Makaiwa
Moalii
Maunalei
Mahana
Paomai
Kaa
HAMAKUAPOKO

is evident in at least one important early cartographic representation of the island. A large map of Hawai'i compiled in 1838 at the Lahainaluna Seminary on Maui was the first to portray a significant number of names of land divisions in the islands (*EMH*, Ill. 66). The section covering Maui is reproduced here. Eleven major land divisions on the island are named using all capital letters and indicated by both coloring and a very fine dotted line for boundaries.

The eastern side of West Maui, while shown in green coloring, is not given a name in all capital letters. The name Wailuku was engraved with an initial capital followed by lowercase letters, implying that it was not considered by the compilers to be at the same administrative level. If that was indeed the case, then Maui consisted of eleven *moku* and an extensive area in the isthmus that was separate but not of *moku* status. Significantly, the coloring of *moku* is basically restricted to an area along the coast. It could be that Samuel P. Kalama, the engraver, and whoever helped him compile the map did not have full knowledge of the boundaries. Or it could be that attempting to show such complex boundaries would have cluttered the map and made the rest of the information hard to read.

By this explanation, there were twenty-nine or thirty formalized *moku* names. Counting the *mokupuni* of Lana'i, Kaho'olawe and Ni'ihau would bring that figure up to thirty-two or thirty-three. Including the *kālana* and the uncertain aspects of Maui's political geography that have been described adds a few more possible major divisions.

There is considerable repetition of *moku* names among the islands, which can be attributed largely to the geographic nature of the names. Kona, which means leeward, for instance, is used on Hawai'i, Moloka'i, O'ahu, and Kaua'i, and always represents a dry area. Ko'olau is the name for windward districts on Kaua'i, Moloka'i, O'ahu, and Maui.

On O'ahu, the *moku* of Wai'anae exhibits a rather unusual feature: A tongue of land stretches over the Wai'anae mountains and runs east across the central plain to the ridge line of the Ko'olau Range. Most explanations for this anomaly generally take the position that the dry Wai'anae coast did not offer a full range of forest products, so the extension to the Ko'olau ridge provided more resources for the people of the Wai'anae *moku*. On Maui, the three *moku* of Honua'ula, Kahikinui, and Hāmākua Loa have long slivers of land reaching to the summit of Pohaku Palaha, the highest point on the east side of the crater of Haleakalā. In these cases, the rationale is more obscure and may be based on something other than the provision of a certain type of resource. Such configurations clearly demonstrate that since Hawaiians did not slavishly define *moku* by simple descriptions, generalizations about the administrative divisions do not always hold true.

Although prominent attention is given to *moku* in early maps and in descriptions of Hawaiian political geography, they figure very little in the *mahele*. When Kauikeaouli divided his land among the various *ali'i*, he did so at the next lower level of administrative unit, the *ahupua'a* and *'ili*. The basic function of the *moku*, it

10 Maui, Lana'i, Moloka'i, and Kaho'olawe, 1838.

Samuel P. Kalama, a noted Hawaiian scholar and government official, made this, the first large printed map of Hawai'i, while at the missionary school at Lahainaluna on Maui. It is notable for the large number of place names that ring the islands perpendicular to the coast, the first instance of a printed map showing many of the *ahupua'a* of the islands.

Kalama, Samuel P. *A map of the Hawaiian Islands according to the latest surveys.* Na Kalama i kaha. Lahainaluna: 1838. Source: Geography and Map Division, Library of Congress; Hawaii—1838—Lahainaluna. Scale ca. 1:590,000. (Note: LC has only two parts of eight, this one illustrated and the title plate.)

11 Eastern O'ahu, 1833

This section of a map drawn by Ursula Emerson (but not surveyed by her) is the earliest known portrayal of *ahupua'a* of O'ahu. Note that there are many *ahupua'a,* indicated by name only, on the well-watered windward coast, but fewer on the dry plain that is the site of modern-day Honolulu.

[Emerson, Ursula.] *Oahu, 1833.* ms. map 59 x 46 cm. scale ca. 1:160,000. Source: Collections of the Hawaiian Historical Society.

appears, was to define a region over which a single *ali'i'ai moku* had control. Within that area, lands were apportioned to various chiefs as custom and political expediency required. It may be that the *moku* reflect the organization of Hawaiian society before the era of frequent warfare, which commenced just before the arrival of Europeans. Future research into Hawaiian political geography may reveal more information about the creation and use of the *moku* concept.

Ahupua'a and 'ili.

The traditional explanation of the Hawaiian land tenure system holds that the *ali'i'ai moku* retained some of the choicest land for himself and distributed the use and control of other lands to lesser *ali'i.* This division was made at the level below *moku* and encompassed two categories of land units known as *ahupua'a* and *'ili*, which figure prominently in the records of the *mahele.*

In traditional Hawaiian society, land and the waters associated with it constituted the sole source of revenue—in the form of food and a limited array of other products—for the *ali'i'ai moku.* He needed to keep enough land for his own use to provide for himself, his family, and the retinue he maintained. To foster political support, however, he also needed to distribute land in a judicious manner to the more powerful *ali'i* beneath him. In theory, the *ali'i'ai moku* had great flexibility in allocating the use of land, but capriciousness was kept in check by common political necessity.

Ali'i were granted large plots of land, equivalent after a

12 Kāne'ohe and Kailua, 1901

The best lands of the *ahupua'a* of Kāne'ohe and Kailua were subdivided into many *'ili*, as shown in this map from O'ahu by John Donn, 1902.

Donn, John. *Oahu, Hawaiian Territory: compiled from all the available data in the office and also from private surveys*. map 101 x 125 cm. Scale 1:60,000. [Honolulu]: Hawaiian Territory Survey, 1902. Source: University of Hawai'i, Hamilton Library, Map Collection, G4382.02 1902 .D718.

13 Taro patches and home, Mauna Wili, O'ahu, 1883.

This area of Maunawili behind Kailua reflects a common situation in the *mahele*. Families often filed a *kuleana* claim for the land they occupied and farmed, which may have consisted of several *apana*, or disconnected pieces, as shown in this photograph. In addition to *lo'i* along the stream, watered by *auwai*, or ditches, and a house site, the claim may have also included groves of banana, hala, or other useful plants.

Source: Bernice Pauai Bishop Museum, Honolulu, Hawai'i; negative number 1-56466, Frame 13, Herman Stolpe.

fashion to the European estate or the New England township. The *ahupua'a* constituted the next division of land lower than the *moku*. *Ahupua'a*, according to Lyons, were defined at the same time as the *moku*.[16] The *'ili* were smaller land enclaves within an *ahupua'a* to which the rights were distinctly separate. Most *ahupua'a* contained no *'ili*, however.

There were approximately eighteen hundred *ahupua'a* in all of Hawai'i and 476 *'ili*. All together, there were about 2,276 major divisions of land available to be parceled out in the islands. The *ahupua'a* were the divisions that contained the necessary resources for a community to prosper. People who lived within an area were well acquainted with the limits of local *ahupua'a*, and those boundaries were defined by natural features, such as a ridge, peak, and shoreline.

The first published map to portray the names of any of the *ahupua'a* was the 1838 map drawn by Samuel P. Kalama, referred to above. Names of *ahupua'a* consist of an initial capital followed by lowercase letters. They are mostly set perpendicular to the coastline, which generally reflects their physical orientation. Kalama relied on the convention of all capital letters for the names of the *moku*, and those names were placed parallel to the coast, thus providing a further visual difference between the two types of land units. No *'ili* appear to be named.

An earlier cartographic portrayal of *ahupua'a* is found on three manuscript maps attributed to the hand of Ursula Emerson, a missionary wife, in 1833 (*EMH* Ills., 61, 63, 64). A portion of one map covering the eastern part of O'ahu has been enlarged here to illustrate some of the physical aspects of *ahupua'a*. In some respects, Emerson's three maps are of great interest and contain a much more complete and accurate geography of those islands than is found on Kalama's map. Manuscript maps that are not shared with others, however, often do not have the lasting impact of printed maps. In that respect, Kalama's map is much more significant. The timing of Kalama's map is of paramount importance to this book, for it was published just seven years before the first steps of the *mahele*. It was the only map that would have been widely available to anyone wanting to study the geography of Hawai'i's *ahupua'a* or to engineer the *mahele*.

Traditionally, *ahupua'a* have been described as encompassing enough land from the mountains to the sea to provide all the resources necessary for the sustenance of a community. In the subsistence economy of Hawaiians, people needed wood and other products from the mountains, water from streams, farmland in the valleys, plains, and lower uplands, and fish from the sea. While the idealized description of an *ahupua'a* as a pie-shaped wedge running from a peak or ridge to the ocean fits many *ahupua'a*, there was actually great variation in the nature of these land divisions. There are a number of *ahupua'a* that are landlocked. Other *ahupua'a* consist of unconnected parts called *lele* or *apana*.

Ahupua'a ranged in size from approximately 180,000 acres to less than ten. According to traditional theory, *ahupua'a* in wetter, more fertile regions tended to be smaller because they provided a greater abundance of foodstuffs and other resources. Close examination of maps showing the boundaries of *ahupua'a*, such as modern topographic maps, reveals that *ahupua'a* size is not explained by that one factor alone, however.

In some ways, it seems appropriate to classify *ahupua'a* into three groups. In one category are large units that truly stretch from high mountain peaks to the sea, and in many cases they mark the boundaries of *moku*. The fact that just a few of these share a common boundary point at a peak or along a ridge means that they effectively cut off the next lower category of *ahupua'a* from the resources of the high mountains.

The next category of moderate-size *ahupua'a* typically encompasses anywhere from a few hundred to a few thousand acres. This type of *ahupua'a* is notable for having access to some limited portion of the uplands, often going part way into a heavily forested area and therefore exhibiting most of the attributes of the idealized *ahupua'a*.

The third type of *ahupua'a* can range from a few hundred to less than ten acres and is remarkable in that it could not conceivably provide all the resources necessary for a community based on a subsistence economy. While some of these *ahupua'a* contained extremely good agricultural land, such as many small units in the vicinity of Lahaina, Maui, others are in leeward areas where agricultural potential was considerably limited. Extensive fishing rights may have gone with these smaller *ahupua'a*, making them a viable unit of society. In many cases, these small *ahupua'a* are completely surrounded by one in the middle-size category, and in other cases they are squeezed in between two much larger units.

The smaller class of *ahupua'a* has many of the physical attributes of the other major category of land unit distributed to *ali'i*, the *'ili*. In general, *'ili* tend to be relatively small, in the range of tens of acres. Although it is a generalization, *'ili* encompass some of the best agricultural land in Hawai'i—or they did before mass urbanization swallowed many of these parcels. Although *'ili* are found throughout the islands, about eighty percent of them are located on O'ahu. The entire *ahupua'a* of Waikiki, from modern-day Ward Avenue to Hawai'i Kai, for instance, was subdivided into approximately ninety *'ili*.

There were two basic types of *'ili*, one that was an integral part of the *ahupua'a* and one that was politically separate. Use of the first type was conferred by the *ali'i* in charge of the *ahupua'a*. It might have been assigned to a favored member of the family or to an important ally of that *ali'i*. According to Pukui and Elbert, this type of land division was known as an *'ili'āina*. J. Cline Mann, a modern surveyor and noted student of the *mahele*, believes this class of land should be called *'ili ahupua'a*.

The second type of *'ili* was politically distinct from the *ahupua'a*, and control of these units came directly from the *ali'i'ai moku*. Pukui and Elbert assign the word *'ili kupono* to this class of land while Mann believes that *'ili'āina* and *'ili kupono* both refer to this type of *'ili*. There are several ways in which the creation of *'ili kupono*, to use the term that the three authorities agree on, might have been of benefit. The creation of a few *'ili kupono* made it possible for the *ali'i* in control of an *ahupua'a* to confer status on several lesser *ali'i*. Creating *'ili kupono* could have served as a way for an *ali'i'ai moku* to reward a greater number of *ali'i* without significantly diminishing the importance of an *ahupua'a* or as a way to establish a significant political presence to counterbalance the power that went with the control of a populous and fertile *ahupua'a*.

Most *'ili* consisted of disjointed pieces of property, in which case it was called an *'ili lele*. In such cases, several plots of land were under the control of a single person even though they might be scattered at some distance from each other within a valley or in different *ahupua'a*. Creating such an *'ili* was a way to tie together several types of land, such as a taro pond, a mulberry plant field, a coconut grove, and a house site. During the *mahele*, these presented a challenge to the surveyors who did not possess a good knowledge of Hawaiian culture.

14 Laupāhoehoe, Hawai'i, ca. 1890

Many *kuleana* claims consisted of irregularly shaped fields, often fenced in by rock walls, which meandered from one corner to the next. Source: Bernice Pauai Bishop Museum, Honolulu, Hawai'i; negative number CP-37536.

15 Grass House, Kaua'i, ca. 1885

An early photograph of a *lo'i* and housesite of a *maka'āinana* family, the basic unit of the original Hawaiian economy. Source: Bernice Pauai Bishop Museum, Honolulu, Hawai'i; negative number CP-79999.

Kīhāpai

Various classes of *ali'i* may have had control of *moku*, *ahupua'a*, and *'ili*, but none of them worked the land and raised the food that fed the whole society. That role fell to the *maka'āinana*, the commoners of Hawai'i. Like the *ali'i* above them, *maka'āinana* were granted the use of particular pieces of land, mostly plots that had long been established and carried with them certain traditional uses.

As can be expected in a language as rich as Hawaiian, there is no single word to describe the plots worked by the *maka'āinana*. *Kīhāpai* is frequently used to describe the smallest parcel of land that was assigned to a Hawaiian family. The root of the word is *hāpai*, which means "to carry" or to "support."[17] From the *kīhāpai*, therefore, the family was able to produce the food that was necessary to sustain it.

The specificity of additional terms reflects the importance of agriculture to the Hawaiian people. *Kō'ele*, for instance, was a plot of land that belonged to an *ali'i* or *ali'i'ai moku* but on which *maka'āinana* were expected to work; since this work was frequently performed on the fifth day of the week, the word for Friday, *Pō'alima*, came to be synonymous with this type of land.[18] Broad paths between taro patches were frequently planted with banana or sugarcane, and these parcels were known as *kuakua*. Another important category was *mo'o'āina*, usually planted in taro. An arable patch of land surrounded by lava was known as a *kīpohopoho*. *Lo'i* were wet taro patches. Other terms exist for lands of similar magnitude but which lack specific definitions.

Kō'ele or *Pō'alima* lands were especially significant during the work of the surveyors of the *mahele*. As will be noted later, disputes arose as to whether certain lands had been worked for the benefit of an individual or as *Pō'alima* lands for the benefit of the *ali'i*.

The preceding synopsis conveys the basic principles of Hawaiian society and land tenure that are necessary to appreciate the surveying work done during the *mahele*. Clearly, not all of the nuances of the Hawaiian land tenure system are completely understood today. Indeed, it is quite likely that many of the men who worked as surveyors during the *mahele* did not understand the system either.

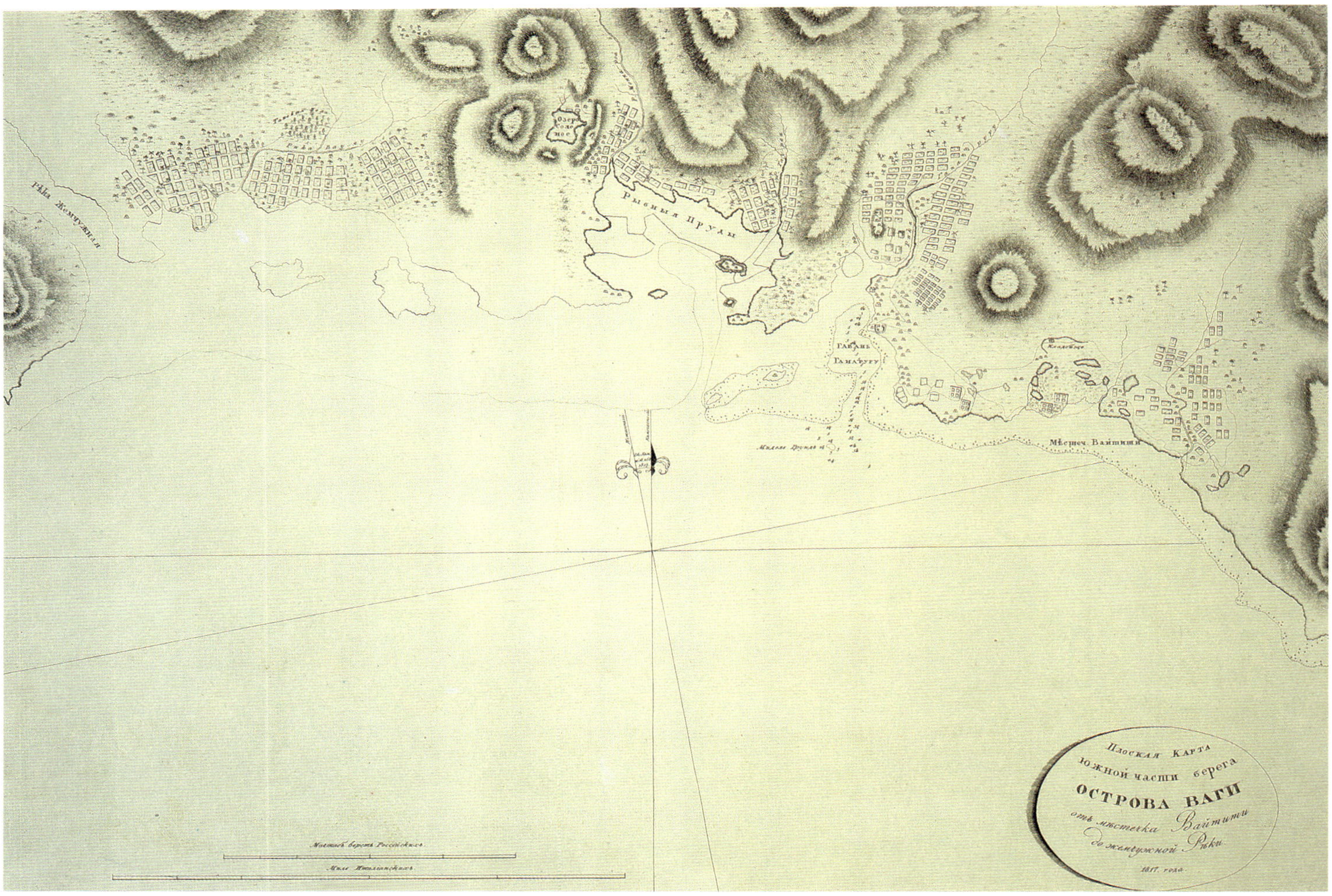

16 South O'ahu, 1816

This map of the area from Waikīkī to Pearl Harbor depicts the extensive irrigated agricultural system developed by Hawaiians under the indigenous land tenure practices. Kotzebue's narrative describes in glowing terms the productivity of this area.

[South O'ahu]. map 46 x 64 cm. Scale ca. 1:30,000. Chart 18 in *Puteshestvie v IUzhnyi Okean i v Beringov proliv dlia ot'iskaniia siervero-vostochnago morskago prokhoda, predpriniatoe v 1815, 1816, 1817 i 1818 godahk. . . . St. Petersburg: N. Grecha. Atlas.* Source: Library of Congress, Rare Book Division, G420.K78 Russian Imperial Collection.

2 Surveying and the Private Ownership of Land

Cartography and surveying share an important quality—they are both as much art as they are science. Knowing which lines to draw and how to draw them requires intelligence, experience, and common sense. While a survey plat is a mathematically based, graphic portrayal of the extent of a piece of property, it is also a complex document that reveals much about the way in which people relate to their land.

Although surveying is essential to private land ownership, surviving documents from the era of the *mahele* do not indicate that any consideration was given to the need for surveying by those advocating this new system. In hindsight, it seems that few foreigners had a thorough understanding of the component parts, including surveying, of this Western notion of private land ownership. In part, this may well have been because of the youth and limited experience of many of those who played crucial roles in the *mahele* process, an issue that is discussed in the following chapter.

The lack of sufficient surveying capability had practical implications in the conduct of the *mahele*. Over a period of about five years, titles were conveyed for approximately 13,000 pieces of property, ranging in size from *ahupua'a* of more than 100,000 acres to *kuleana* of less than an acre and taro patches of less than one-tenth of an acre. Yet for none of these was there anything vaguely resembling a written description appropriate to the Western legal system that Hawai'i would soon adopt. In fact, at the start of the *mahele* there was only one man in Hawai'i who was a trained and experienced surveyor, Theophilus Metcalf, and only two with equal qualifications, William Webster and Rudolph Meyer, who came later. A few missionaries used their textbook knowledge of the discipline to practice surveying and teach it to others, but they had no extensive practical experience.

Given the scarcity of surveying practitioners, as well as other professionals and government institutions, why did the foreigners occupying positions of trust, responsible for teaching and advising the Hawaiian royalty, advocate private land ownership? For the most part, their motivations are best left to historians who approach this era from a much broader perspective than is found in this work. But an examination of mapmaking and surveying adds new insight to this discussion.

Statements by many visiting and resident foreigners reveal their implicit faith in the moral and economic superiority of private land ownership. Yet private ownership of land is not universally practiced. Nor is it a single idea. Even in Western societies, the widespread practice of land ownership as it is now known is only a few centuries old. The practices of land tenure in feudal Europe, for instance, differed significantly from those in Europe and America in the mid-1800s. The concept of extensive private land ownership is still a new idea in Russia, in many of the other former Soviet states, and in numerous "semi-Western" societies. The variety of land tenure patterns found throughout the world historically and today shows that humans have developed many different ways of allocating the use of land.

The views of foreigners in Hawai'i on the virtues of private land ownership have been well-documented in other sources. Yet the comments of three—a foreign official, a resident businessman, and an American missionary—provide noteworthy evidence of the way cultural beliefs and biases can be reflected in something as seemingly straightforward as the concept of private land.

George Anson Lord Byron, commander of the H.M.S. *Blonde*, brought home to Hawai'i the bodies of Liholiho, Kamehameha's successor, and his wife Kamāmalu, who died of measles in June 1824, just two months after arriving in London on a diplomatic mission. Considering that Byron was under explicit instructions to be gracious and helpful to the people of Hawai'i, his remarks represent at least conventional wisdom if not official British policy. In the official narrative published about the voyage, he is quoted as advising the Hawaiians that

> ... the lands which are now held by the chiefs shall not be taken from them, but shall descend to their legitimate children, except in the cases of rebellion, and then all their property shall be forfeited to the King.[1]

Byron's remarks clearly espouse the Western concept of private land that is passed freely from one generation to the next. Implicit in his remarks, however, are other cultural beliefs intertwined with land ownership. First, his recommendation that land be confiscated in the case of rebellion places the ownership of land subordinate to loyalty to the king or government. The punishment for a man's transgression against the highest authority would thus justifiably be extended to his descendants. This link between land ownership and deference to the government would have benefitted Europeans and Americans, whose dependence on Hawaiian ports made them anxious to ensure domestic tranquility.

Another aspect of Western culture inherent in Byron's comments was his reference to "legitimate" children. Although land was theoretically private in England and elsewhere, aspects of land practice effectively helped enforce social customs, with laws regarding primogeniture and the lack of legal standing for illegitimate children being prime examples. Applying Byron's standard of "legitimacy" would have been difficult in Hawai'i, where polygamy existed and a broad concept of family prevailed. In a society where social standing, or *mana*, was determined through a very complicated process of genealogy, there was no easy analogy to the concept of "legitimate" birth.

Byron's remarks, therefore, illustrate how many ideas of culture are ingrained in land tenure practices. While Byron may have thought his advice to be perfectly reasonable, the implementation of his suggestions would have required radical changes in Hawaiian society.

Another Englishman, the acting British consul general to Hawai'i, offered similar advice but extended it to the role of government. In a memorandum of September 25, 1831, to Kauikeaouli, William Miller advised the young *ali'i 'ai moku* that there should be

> some defined form of government, and a few fundamental laws that will afford security for property.... [2]

The role of this government, he noted,

> ought to be to ameliorate the condition of the people—to do away gradually with the actual despotic feudal system, and thus convert the condition of the degraded serfs, as the kanakas now may be styled, into industrious and respectable people.[3]

Miller clearly believed that providing security for property, another way of advocating private ownership of land, was as essential as codifying the form of government. Would land ownership, along with a new form of government, transform the Hawaiians "into industrious and respectable people?" The more important question, though, is whether Hawaiians were not "industrious and respectable people?" At least some foreign visitors—as is noted later in this chapter—seemed to disagree with Miller's assumption, and the nature of their comments has bearing on the issue of land use and the subsequent surveying of the *mahele*. In Miller's defense, however, it should be noted that his remarks were penned during the period when the impact of gathering sandalwood was having a significant detrimental effect on the Hawaiian economy and society.

One aspect of Miller's remarks, however, should be emphasized: his recommendation "to do away *gradually* [emphasis added] with the actual despotic feudal system. ..." At least he recognized the importance of instituting change slowly. His advice seems to have been lost on those who were pushing for private ownership in the 1840s, however, for the *mahele* was not a gradual process given the vastness of the enterprise.

Typical of the advice given Kauikeaouli and other Hawaiians by missionaries is found in the statement of John Emerson, the missionary at Waialua, O'ahu, who wrote that

> I think two things are requisite to make these people industrious and provident. First, the feeling that the land they cultivate is their own, for themselves and their posterity. Second, the feeling that this land is of real value, and capable of being improved in value, and that all improvements are private gain.... [4]

He added,

> In my opinion, the people must own their lands and feel at liberty to sell them for gain to themselves, or the Hawaiian race will never rise to the habits of industry.[5]

Many similar comments on the relative merits of Hawaiian land tenure and private land ownership indicate how common such opinions were among foreigners. But those with an open mind could find something of value in the Hawaiian relationship to the land. The comments of Otto von Kotzebue, a Russian naval officer who visited the islands in 1817 and 1818, negate some of the criticism implicit in those statements that relate industriousness to the private ownership of land. Kotzebue's flowery description of the landscape of southern O'ahu implies that without the benefit of private land ownership the Hawaiians created an agricultural system equal to any in the world in terms of organization and productivity. Of O'ahu, Kotzebue noted that

> it is called the garden of the Sandwich islands, and it has a right to this name, on account of its extraordinary high state of cultivation. . . .[6]

He describes this system of cultivation in some detail:

> The artificial taro fields, which may justly be called taro lakes, excited my attention. Each of them forms a regular square of 160 feet, and is enclosed with stone all round like our basins. This field, or rather this pond, such it really may be called, contains two feet of water in the slimy bottom of which the taro is planted, as it does not thrive except in such a wet situation: each pond has two sluices, to let in the water on one side, and out again at the other, into the next field, and so on. The fields are gradually lower, and the same water which led from an elevated spring or rivulet, can water a large plantation.[7]

Kotzebue continues,

> I have seen whole mountains covered with such fields, through which the water gradually flowed; each sluice formed a small cascade, which ran through avenues of sugar-cane or banana, into the next pond, and afforded an extremely picturesque prospect.[8]

Kotzebue's observations are echoed by Jacques Arago, an artist and writer who visited Hawai'i aboard the French vessel *L'Uranie* in 1819:

> The environs of Lahaina [Mau'i] are like a garden. It would be difficult to find a soil more fertile, or a people who could turn it to greater advantage.[9]

This complex system of agriculture was built before the foreigners arrived. It required industrious men—and in the traditional Hawaiian society it was the men who did this work—to construct it, and it required attentive men to keep it working. Irrigated systems such as the Hawaiians developed do not last long if left unattended or if managed improperly. A walk today through one of the abandoned valleys, such as the densely overgrown Halawa on Moloka'i, illustrates what happened once the agricultural landscape was left unattended. The existence of such a well-developed agrarian structure demonstrates that under the original land tenure system, Hawaiians were industrious enough to build and maintain this highly productive system, and they did so without the motivation of private ownership to secure their long-term interests.

While the comments of observers like Kotzebue and Arago tend to negate criticism from Byron, Miller, Emerson and others, it must be remembered that all aspects of Hawaiian society suffered tremendous convulsions in the 1820s, 1830s, and 1840s in addition to the catastrophic loss of population that followed the arrival of foreigners. So a system that seemed superlative in the eyes of viewers just before 1820 was exhibiting signs of disintegration over the next two or three decades.

Nonetheless, many foreigners did express a strong belief in the intellectual and cultural superiority of private land ownership, and since Hawaiians were embracing many aspects of Western economic and social belief, it must have seemed logical to foreigners to suggest that Hawaiians endorse private ownership of land as well.

The complexity of private land ownership raises another important issue: Many cultural institutions are necessary to support private land ownership in the Western sense. In most cases, these cultural institutions take on legal or quasi-legal form. Needed are agencies for the registry of deeds and titles, a body of law describing the rights and restrictions imposed on ownership and transfer of land, a judicial system with a body of case law or precedence for handling disputes, attorneys and other professionals who prepare and examine legal documents related to land transactions, and policies that clearly define the role of land in raising revenues for the operation of the government.

Such institutions are required in Western society because of the multiplicity of land rights. Out of the feudal system of medieval Europe evolved conventions governing the use of land within the framework of private ownership. Instead of a single category of land rights, there are several classes of land, some burdened with greater restrictions than fee simple, such as the renting or leasing of property and the buildings on it. With these classes there are a variety of conditions, such as the right to sublease or otherwise pass on the use of rental property. In contemporary Hawai'i, for instance, many people own their homes but lease the land from an individual, private company, or estate. The concept of private land is still evolving; in recent decades, new categories have been created or become more popular, such as owning a specific piece of land and having limited rights to the use of additional communal land, as in a townhouse setting. And "property" rights now pertain even to the ethereal, as in the ownership of "air rights" in a multi-level commercial or residential condominium and to temporal rights as in time-share properties.

Specialists with more expertise in professions related to land ownership may better comment on how well foreigners had prepared the Hawaiians for the complex system they advocated so strongly. From the perspective of cartography and surveying, however, it appears that little concern was given to the surveying capacity that existed to document the transformation of the Hawaiian land tenure system.

For this volume, a significant issue is whether ownership practices in Western society were so ingrained in people of European and American descent that their perception of the Hawaiian land tenure system was subconsciously altered. The role of the surveyor is to describe in unambiguous terms the location and extent of a piece of property. This involves stating the configuration of its boundaries and its position in relation to other features in a well-defined frame of reference. The surveying work of the *mahele* involved interpreting the existing patterns of land use and describing them in the symbology of an alien culture. The surveyors' contribution to land ownership is a type of map known as a plat.

To appreciate the relationship of surveying to land ownership, one must recognize that the apparently simple elements comprising a plat are weighted with significant cultural meaning. In fact, three specific basic assumptions underlie the discipline: that surveys are forms of communication; that complex landscapes and environments can be represented by simple symbols; and that a line divides space.

Drawing a map of any type, from a world map to a survey plat, is a form of communication fraught with all the potential problems of any means of speech or art. In recent years, scholars working on the history of cartography have taken care to note that all cultures do not view maps the same way. Maps are abstractions of the real world, even when drawn to scale and when all features are placed according to precise measurements. The mapmaker must decide which features to portray and what type of symbology to use, decisions influenced by that person's experience, education, culture, and religion. This abstraction process is so shaped by past practices that cartographers—and surveyors—often use common conventions without even thinking about them.

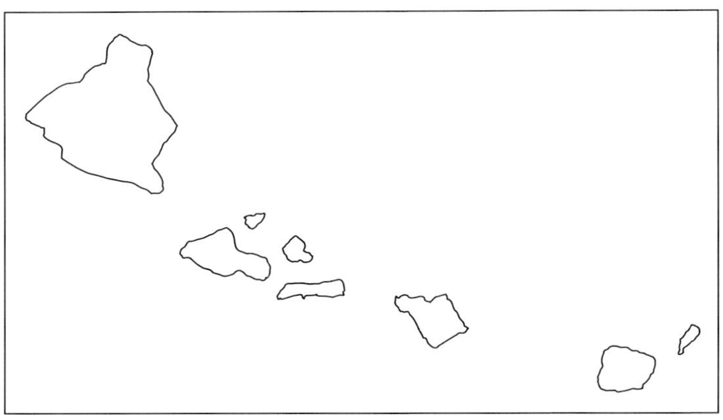

This outline map of Hawai'i illustrates one such convention: On most maps in Western society, North is oriented to the top of the page. A map of a normally familiar place that is oriented differently can be disconcerting. Just as there is no universal orientation of writing in various cultures, there is no single way to draw maps and survey plats. Since cartographic images are a form of communication, practitioners logically use symbols and techniques that the intended audience will recognize. But there is no guarantee that a viewer from a different culture will interpret the symbology and conventions in the way the mapmaker or surveyor intended.

The particulars of such cartographic conventions are not as significant as the fact that they are a component of communications within a cultural group. The surveyors of the *mahele* were often faced with interpreting what parts of a local landscape belonged within the area claimed by an individual. This frequently involved extended discussions between a surveyor less than fluent in Hawaiian and Hawaiians who were not conversant in English. This was more than a language problem, however, for it involved the translation of radically different concepts of land use. The difficulties were compounded by the fact that the surveyor used a limited set of conventions and symbology, which in turn influenced his view of what he saw.

A second significant point involves the assumption that complex geographic features can be represented by relatively simple symbols. If the above illustration is rotated to its "correct" orientation, readers will recognize it as an outline map of the islands of Hawai'i.

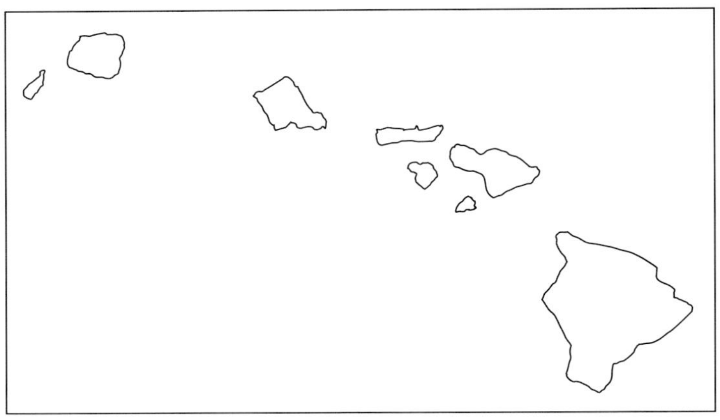

How can eight squiggly lines represent something as complex as an archipelago? The answer lies in part with the familiarity of the image, but also in a learned reasoning process: Areas enclosed by a line are assumed to be different from the areas outside. If a person knows that Hawai'i consists of islands, then it follows that land lies within the lines, water on the outside. The mind is willing to assume that the line symbolizes the coastline of an island.

Humans are able to interpret the abstraction of complex entities as long as they have a basic familiarity with the conventions being used. Such conventions can become habit-forming. Pages of this volume could be filled with a listing of the multitude of natural and manmade features that are routinely symbolized through the use of a line: a coastline, a stream, a road, a boundary, a power line, etc.

Another aspect of cartographic symbology comes into play in portrayals of landscapes: generalization. Here the issue is the kind of line used to represent a geographic feature. Should an island, for instance, be represented by a single curved line or a chain of straight line segments? Generalization is more acceptable in some contexts than in others. The illustration below contains three cartographic representations of O‘ahu.

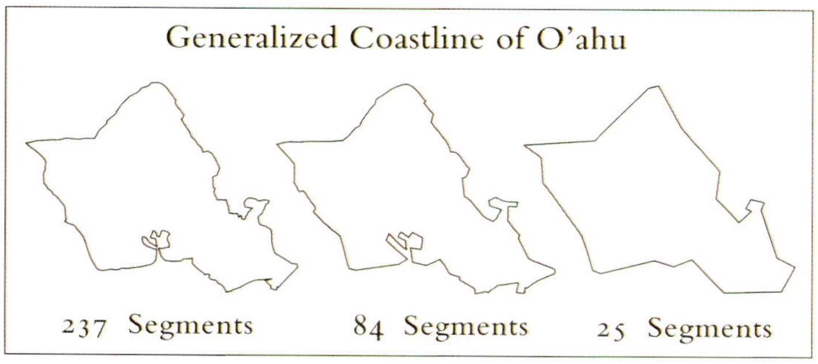

O‘ahu's 112 miles of coastline are represented by 237, 84, and 25 line segments respectively. In different situations, any of these representations may be considered acceptable.

Clearly, surveyors use the same form of abstraction. A rock wall separating two properties may not be perfectly straight, but a surveyor may choose to treat it as such so that it can be portrayed as a single line rather than dozens or hundreds of smaller segments. A ridge line between two *ahupua‘a* may appear on a survey as a few straight line segments rather than as hundreds of irregularly shaped segments.

In the field, surveyors of the *mahele* may have faced great constraints as they tried to view landscapes, such as land districts or *kuleana* lands, in ways that could be mapped by the set of symbols they were trained to use.

An issue of symbology very germane to the work of surveyors is the meaning of a line. Conceptually, a line on a map either connects points of equal value, as in a contour line or a road, or separates things that are different, as in a boundary line. In surveying, lines are generally used to separate different entities, such as the two irregularly shaped parcels that share a common border, shown below.

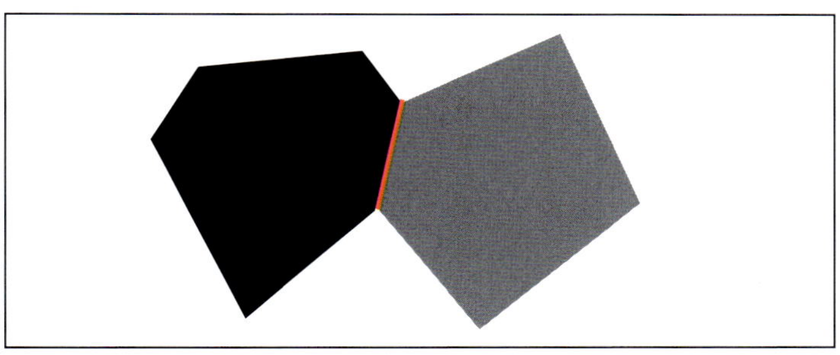

The red line implies that everything to one side is different from that on the other side. The line is infinitely small in concept and actually represents no actual space on the face of the earth. It is simply an expression of the abstract belief that all land is either part of the parcel on the right or the one on the left.

Such lines are found on every survey drawn as part of the *mahele*. Yet the use of such a line may not have been appropriate in some scenarios. Did the Hawaiians, for instance, assume that the division between adjacent *kīhāpai* occurs in the middle of walls that separated them? Or was the wall assumed to belong to one parcel, and its legitimate owner obligated for its maintenance? Or was it considered a part of neither parcel, and the *kīhāpai* viewed as only that land within the walls?

In the first chapter, the term *kuakua* was defined as a cultivated path between two fields on which fruit was grown. Did a *kuakua* belong to one of the fields, was it shared by the users of the adjoining fields, or did it have a communal value in the sense of a pathway or right-of-way? The surveyors of the *mahele* had to first recognize such unique aspects of the Hawaiian land tenure system, and then describe them within the conventions and symbology of a different culture that did not always offer direct analogies.

What about streams and *‘auwai*, or irrigation ditches? Hawaiians had well-defined conventions for the use of water, known as riparian rights in Western terminology. When surveyors indicated the boundary of a *kuleana* running along a stream, did they consider the boundary to be at the edge of the stream or down its middle? Such questions bear on the use of private lands. As long as Hawaiians were alive who understood all the rights associated with particular pieces of property, such issues could be settled amicably. But the ambiguity resulting from generalizing such features can be the source of great controversy when only the surveyor's plat and notes exist to describe a piece of property that was used by Hawaiians according to age-old traditions.

Although the surveyors of the *mahele* were supposed to describe existing components of the Hawaiian land tenure system, in fact they helped redefine it in ways consistent with the symbology of Western surveying practice. Evidence of the original land tenure system is present in the surveys of the *mahele* era, but any interpretation of the survey records and plats must be made with an understanding of two factors: the relationship of surveying to the concept of private land ownership and the ways in which surveying practice may have influenced surveyors' perceptions of the lands they were charged with describing.

The Mechanics of Surveying

That aspect of surveying concerned with the description of property is generally referred to as land surveying. To cite a textbook, this subdiscipline focuses on

> the location of property boundaries and the preparation of drawings (or plats) which show these boundaries. In addition, it involves the writing and interpretation of land descriptions involved in legal documents for land sales or leases.[10]

Today, much of the work of land surveyors involves re-establishing the boundaries of existing properties so that all parties involved in a transaction clearly understand the extent of the plot of land at issue. In the time of the *mahele*, surveyors faced a different issue, defining for the first time the boundaries of plots being transferred to private ownership.

In land surveying, the measurement of lines and angles is only a small part of the job. Just as important is knowledge of the law regarding land and past practice in a given area. Modern property law in Hawai'i, as in all states except Louisiana, traces its roots to English common law, which dates back to 1272. Since most of the surveyors working in Hawai'i received training in English and American practice, the importance of English common law and the practices that it embodies can be seen as crucial in interpreting the surveying work done during the *mahele*.

When surveyors complete their interpretation of the nature of a piece of property and determine the extent of its boundaries, whether it is original or resurveying work, they make a plat, or scaled drawing, of their findings. A subsequent surveyor could return to the site, start at any of the identifiable points, measure the appropriate angles and lengths of lines, and end up exactly at the starting point. For that to be possible, however, the original surveyor would have had to give at least two pieces of essential information: how the starting point was found and the frame of reference by which the angular measurements were determined.

Where surveying is well-established, conventions and regulations dictate how the work is to be done and documented. But in Hawai'i during the *mahele*, there was no local tradition. Surveyors came from a variety of backgrounds, and some of them could hardly be called professionals. Consequently, many of the *kuleana* surveys provide no clear indication where property lies in relation to some well-known and permanent frame of reference.

The measurement of the lengths of boundaries is important for two reasons. First, that information is essential for determining the size of the property. An infinite number of polygons have the same angular measurements. By providing lengths of the sides, the figure is given a scale which is then used for computing the area of the property. Second, the lengths of the sides provide an accuracy check for the survey. Only one possible polygon has a given set of angles and a given set of side lengths. In theory, only one side length is needed to describe a piece of property if all the angles are known, but knowing the length of all sides makes it possible to determine whether previous surveyors made errors in their measurements of angles or lengths.

Measuring the length of a side is not a simple process. Consider the following cross-section of the 200-foot boundary from the above illustration.

Assume that the boundary runs from the tree at the bottom of the hill to the large stone located higher on the slope. In the following illustration, different colors reflect

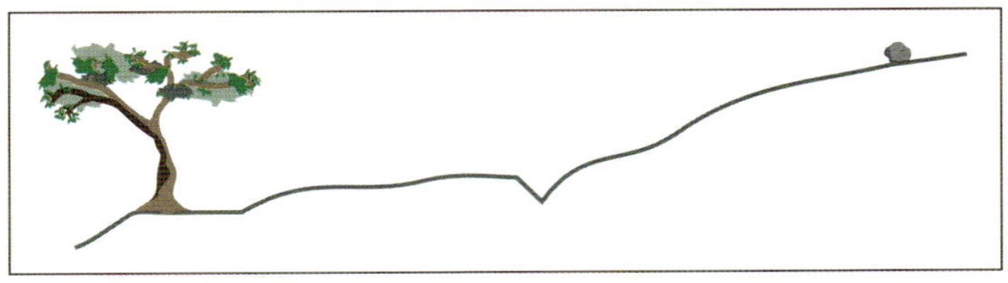

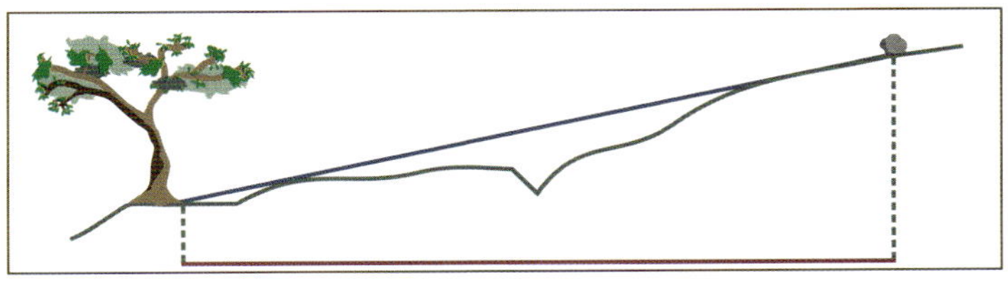

three ways to measure the length of the boundary segment.

The green line represents a measuring device laid along the ground. At the time of the *mahele*, a surveyors chain would have been used, which consisted of 100 links that yielded a length of 66 feet. The blue line represents a hypothetical line pulled taut between the tree and the large stone. The red line represents the line traditionally represented on a survey plat, the horizontal distance between the two end points that ignores any difference in elevation.

This convention of using the horizontal measurement came about because it provides the only means for determining the location of boundary corners on a variable landscape. In the above illustration, it is possible that adjoining owners could have filled-in the slight gully in the middle of the boundary to control drainage, thus changing the distance when measured directly on the ground. A slight subsidence at the upper part of the slope due to changes in the water table would alter the "taut line" distance between the two ends of the section. By always referring to the horizontal distance between two points assumed to be on the same plane, the limits of the property can be determined regardless of future changes.

In practice it is often necessary to cut vegetation in order to gain a clear sighting between two points. To get surveying instruments to one of the turning points may entail difficult climbing. Temporary shelter may have to be erected to keep rain and wind from affecting the instruments. For professionals, such occupational hazards are part of the job. In Hawai'i during the *mahele*, however, anecdotal evidence suggests that many of the less than professional surveyors took short cuts to alleviate some of the tedium and difficulty associated with the work.

When fields were separated by such features as *kuakua*, much of a property line could have been obscured by banana plants or sugar cane. Cutting away vegetation to get a clear line of sight was not only difficult, it would have caused economic hardship. Consequently, the surveyor was forced to find an alternate method of determining the angle and distance of that segment. It could have been done as simply as moving the transit and the rodman's pole an equal distance off the line, or it could have involved measuring all the other angles and lines in the property and then determining the angle and distance of the missing line from the scaled drawing. In a society where the practice of surveying is well-regulated, there are accepted ways for accommodating such situations. In Hawai'i, surveyors likely employed many undocumented shortcuts that greatly reduced the accuracy of their work.

Consider the task of surveying a whole *ahupua'a*. In windward areas, the ridge lines that often separated *ahupua'a* were heavily vegetated, slippery from the heavy rainfall, and bounded by exceedingly steep slopes. Imagine carrying surveying instruments high into the mountains and then dragging a surveyors chain along a slippery ridge barely wide enough to walk. The temptation to find an easier way was clearly present.

One solution, called triangulation, was to measure ridge lines from a much safer vantage point. To do this, a surveyor would mark two points in a relatively convenient location. The distance between the two points would be carefully measured, forming a baseline. Then, from both end points the surveyor would record the angular bearing to major identifiable points on a ridge line. On a sheet of paper, he would first draw the base line to scale. Then, for each landmark on the ridgeline he would draw a ray from each end of the baseline according to the angles he had measured. The intersection of the two lines marked the location of the point. Using dividers (or geometry) and a protractor, he would then calculate what the angles and distances should be and write the survey description as if he had made those measurements directly with his transit and chain.

There are other ways surveyors cut corners when the situation in the field proved difficult or insurmountable. The classic example noted about surveyors in the *mahele* is described by W. E. Wall, the director of the Hawaiian Territorial Survey from 1901 to 1929:

> It may serve to illustrate some of the old methods of surveying to cite the case of one who had been an old surveyor of public lands in the United States and who, when confronted with the dense jungles of Hawaii, was accustomed to send an assistant far ahead, out of sight and nearly out of hearing, and then sight his instrument as best he might at the assistant's "yell," borne faintly on the breeze, after which two chainmen, by any route that they found the easiest, would measure a distance that was recorded with microscopic precision.[11]

Shortcuts such as these certainly lessened the reliability of surveying during the *mahele* but were sometimes necessary or expedient. Had the surveyors documented their work more completely, subsequent surveyors and courts could have given due weight to these factors. Unfortunately, not only did surveyors frequently fail to record these shortcuts, they often failed to indicate such basic facts as whether they adjusted for variation of the magnetic compass.

One of the biggest problems in the *mahele*, however, was the failure to provide a frame of reference as to the starting point of a survey. This was largely true for the survey of *kuleana* grants to *maka'āinana*. To some extent, this was not the fault of the surveyor but represented the fact that there was no infrastructure of clearly defined locations to which surveys could be referenced.

Without some clear reference to the approximate location of a piece of property, accurately measured angles and lines are meaningless. Traditionally, this was accounted for by providing the names of adjoining property owners or land names. If the extent of all surrounding properties are known, then a survey can be recreated by referring to the records of those surveys. During the *mahele*, however, many *kuleana* grants were islands within an *'ili* or *ahupua'a*; there were no neighboring parcels to reference.

As a result of this failure to reference control points, individual properties could be lost over time. If a family rented its property to a nearby landowner, such as someone developing a sugar plantation, the trees, walls, ditches, and other natural features that defined it might well have been destroyed in the effort to produce larger, more efficient field patterns. When a subsequent heir tried to relocate the property, the identifying features of the boundaries were gone and there was no well-established control station from which the starting point of the survey could be determined.

As the above discussion illustrates, there were many ways in which a surveyor of the *mahele* era could have produced faulty work. It should be noted, however, that an erroneous survey does not have legal standing if the original property boundaries can be clearly identified. If, for instance, a survey described a boundary as being along a stone wall but the angular and distance measurements on the plat mistakenly place it fifty feet into a neighbor's land upon resurveying, the wall is still the boundary. A surveyor's mathematical mistakes or errors of observation do not take precedence over the common understanding of property boundaries. But over time, people die and landscapes are changed. If there are no clear indications of where a boundary was—if the stone wall had been removed—and the owners of the two properties disagreed over the boundary line, then the survey plat takes on more significance because it is the only surviving record from the time when the boundaries were well-known.

The survey work of the *mahele*, despite the poor quality of much of it, became unduly important as knowledge of property boundaries and the uses of land was lost because the Hawaiian population continued to decline and because significant numbers of Hawaiians were moving away from the land and into the towns and villages that were appearing throughout the islands. The surveys made during the *mahele* came to represent the memory of much of the landscape of the traditional Hawaiian tenure system.

17 Kauikeaouli (1813–1854) ca. 1850

Also known as Kamehameha III, Kauikeaouli occupied the throne during the transformation of the Hawaiian land tenure system known as the *mahele*. Source: Bernice Pauai Bishop Museum, Honolulu, Hawai'i; negative number CA-1041.

3 The Mechanics of the *Mahele*

Of all the cultural and political changes that swept through Hawai'i in the 1830s, 1840s, and 1850s, the *mahele* was probably the most significant. The *mahele*, however, needs to be viewed as part of an overall effort to create a formal government structure in the Western sense with the creation of a constitutional government, the establishment of government ministries, and the writing of a broad range of basic legal statutes.

Among the many significant cultural changes of that era were several events with some bearing on the *mahele*. To understand such events and the ambiance of the time, it is important to remember that Kauikeaouli was only twelve years old when he assumed the crown in 1825 as Kamehameha III. Since he was too young to rule, Ka'ahumanu, who had been serving as *kuhina nui* since 1823, was appointed regent. A childless wife of Kamehameha I, Ka'ahumanu was the defacto ruler during her nine-year tenure. The *mahele* confirmed many awards of land made by Ka'ahumanu during her regency. After her death in 1832, the regency fell to the young king's half-sister, Kīna'u.

It is well-documented that most of the *ali'i* in Hawai'i sought to solidify their status and power during Kauikeaouli's youth. Certain elements of the upper echelon—including Ka'ahumanu—hoped that Kauikeaouli would never exercise the power possessed by his late father, Kamehameha. The establishment of a hereditary monarchy, which was favored by foreigners, would formalize a political system that would leave many of them on the fringes of power. As he matured, therefore, the young man was confronted with difficult challenges on two fronts—the political machinations of the *ali'i* and the encroachment of foreign culture. As explained in Lilikalā Kame'elehiwa's work, *Native Land and Foreign Desires*, Kauikeaouli's ultimate decision to implement private ownership of land was greatly affected by these two powerful forces.

The story of Kauikeaouli's governance is far too complex to treat in depth here, but it can be stated that he failed to preserve the authority and power of the position that his father had created through his conquest of the islands. Although his reign is generally described in very favorable terms, Kauikeaouli gradually gave away much of his power to various other elements of Hawaiian society. In the end, he was no longer the *ali'i 'ai moku*, but rather the *mō'ī*, or monarch, who exercised authority within the confines of a constitutional system modeled on Western political structures.

The first formal step along the path toward private land ownership was the Bill of Rights of 1839. This important act actually changed little, for it largely codified existing practice. The fact that the rights of individuals were set down on paper, however, was in itself an important alteration in the nature of Hawaiian society.

One facet of the Bill of Rights of 1839 is worth noting here: It described certain rights the *maka'āinana* had regarding their land:

> Protection is hereby secured to the persons of all the people, together with their lands, their building lots and all their property, and nothing whatever shall be taken from any individual, except by express provision of the laws.[1]

There was also some recognition of the right of inheritance involving small portions of land, but with limitations:

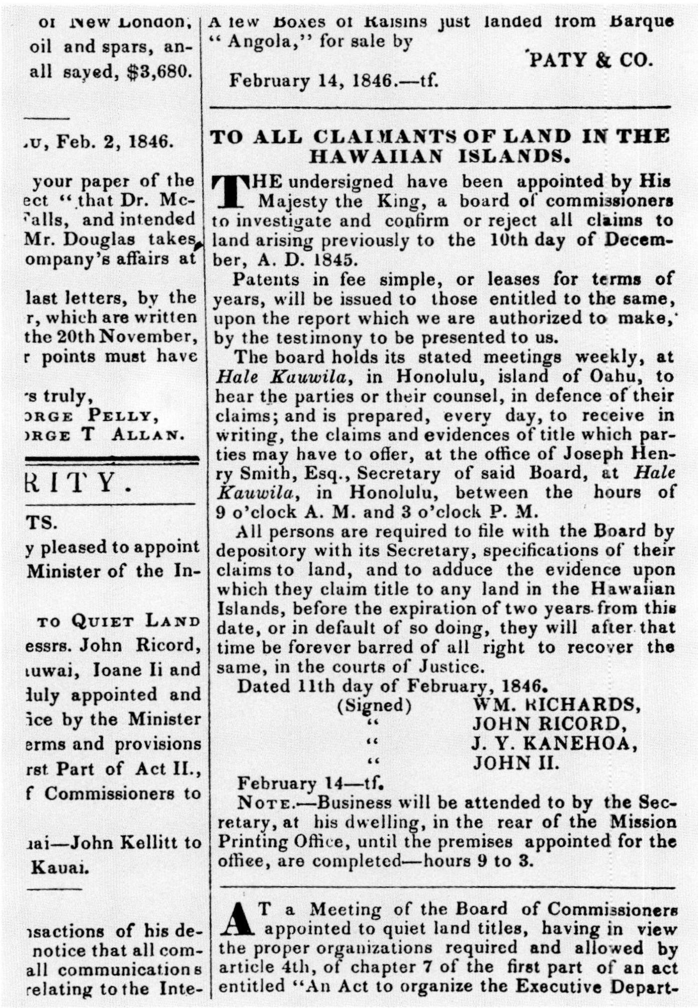

oi New London, oil and spars, an- all sayed, $3,680.

.u, Feb. 2, 1846.

your paper of the ect "that Dr. Mc- 'alls, and intended Mr. Douglas takes ompany's affairs at

last letters, by the r, which are written the 20th November, r points must have

's truly,)RGE PELLY,)RGE T ALLAN.

RITY.

TS.

y pleased to appoint Minister of the In-

TO QUIET LAND essrs. John Ricord, uwai, Ioane Ii and duly appointed and ice by the Minister erms and provisions rst Part of Act II., f Commissioners to

iai—John Kellitt to Kauai.

isactions of his de- notice that all com- all communications relating to the Inte-

A few Boxes of Raisins just landed from Barque "Angola," for sale by PATY & CO.

February 14, 1846.—tf.

TO ALL CLAIMANTS OF LAND IN THE HAWAIIAN ISLANDS.

THE undersigned have been appointed by His Majesty the King, a board of commissioners to investigate and confirm or reject all claims to land arising previously to the 10th day of December, A. D. 1845.

Patents in fee simple, or leases for terms of years, will be issued to those entitled to the same, upon the report which we are authorized to make, by the testimony to be presented to us.

The board holds its stated meetings weekly, at *Hale Kauwila*, in Honolulu, island of Oahu, to hear the parties or their counsel, in defence of their claims; and is prepared, every day, to receive in writing, the claims and evidences of title which parties may have to offer, at the office of Joseph Henry Smith, Esq., Secretary of said Board, at *Hale Kauwila*, in Honolulu, between the hours of 9 o'clock A. M. and 3 o'clock P. M.

All persons are required to file with the Board by depository with its Secretary, specifications of their claims to land, and to adduce the evidence upon which they claim title to any land in the Hawaiian Islands, before the expiration of two years from this date, or in default of so doing, they will after that time be forever barred of all right to recover the same, in the courts of Justice.

Dated 11th day of February, 1846.

(Signed) WM. RICHARDS,
" JOHN RICORD,
" J. Y. KANEHOA,
" JOHN II.

February 14—tf.

NOTE.—Business will be attended to by the Secretary, at his dwelling, in the rear of the Mission Printing Office, until the premises appointed for the office, are completed—hours 9 to 3.

AT a Meeting of the Board of Commissioners appointed to quiet land titles, having in view the proper organizations required and allowed by article 4th, of chapter 7 of the first part of an act entitled "An Act to organize the Executive Depart-

18 The Polynesian, February 14, 1846

This notice in the major newspaper of the islands gave the Hawaiian people two years in which to file claims for land they had been farming. It was reprinted often in subsequent issues.

Source: Hawaiian Mission Children's Society, Honolulu, Hawai'i; negative number N-1843.

> But those individuals who possess three or more divisions of land can bequeath only two-thirds to heirs, and the other third reverts to the King.[2]

This latter provision relates only to the *ali'i*, for the *maka'āinana* certainly did not possess whole divisions of land. In one respect, it might be argued that the Bill of Rights did not actually give the *maka'āinana* the right to own land, only that due process was required before they could be evicted. It did nothing to change the fact that *maka'āinana* and the *ali'i* still held land by the grace of the king.

In 1840, Kauikeaouli, then only 26, agreed to a more far-reaching document, one that instituted a constitutional monarchy. While the most important aspects of the Constitution of 1840 provided for the division of power among the *mō'ī*, the *ali'i*, and the *maka'āinana*, it indirectly addressed the land issue as well. In a statement regarding the "origin of the present government," the Constitution noted that

> Kamehameha I, was the founder of the kingdom, and to him belonged all the land from one end of the Islands to the other, though it was not his own private property. It belonged to the chiefs and people in common, of whom Kamehameha I was the head, and had the management of the landed property.[3]

While affirming that all land belonged to Kauikeaouli, the Constitution also decreed that it was not his personal property. He was merely caretaker of the land in the name of the *ali'i* and *maka'āinana*.

One reason behind Kauikeaouli's ultimate decision to accede to foreign suggestions and institute private property rights is that he was clearly aware that his country was vulnerable to the imperialist ambitions of a European nation or the United States. In fact, France colonized Tahiti and the Marquesas Islands in the early 1840s.

Aggressive action in 1839 by a French ship and then in 1843 by a British man-of-war jeopardized Hawaiian independence. In both instances, higher authorities had overridden the actions of the naval captains, and Hawai'i had retained its sovereignty. But Kauikeaouli knew that he might not be so lucky in the future.

One way for Kauikeaouli to protect himself—as opposed to protecting the monarchy—in the event of foreign takeover was by instituting private land ownership. The Constitution of 1840 stated clearly that Kauikeaouli had no right to land in his own name; it all belonged to the government. If the government fell to a foreign power, so too did his right to land and the proceeds from it. Many foreigners, especially the missionaries, pointed out to Kauikeaouli that by granting to others the right to own land, he also would give himself the same privilege. In other words, he would be able to assign lands to his personal ownership instead of having everything belong to the government.

Part of the goal of the *mahele*, therefore, was to separate the government from the person of the *mō'ī*. As *mō'ī*, Kauikeaouli would have control of the government's land as well as that which he would designate to be his private property.

The Land Commission and the Mahele of 1848

With many concerns on his mind, Kauikeaouli embarked on the irrevocable course of implementing private ownership of land. To understand what transpired over the next few years, it should be remembered that the *mahele* evolved as it progressed. Hindsight reveals that there was no masterplan or timetable in which all the fine details of what was to happen were mapped out. Consequently, some statements of principles seem not to have been executed, and some legal acts seem to contradict prior ones. Many of these inconsistencies cannot be clarified with certainty. Debate still exists over specific wording of documents or the intentions of the individuals involved in the *mahele* just as debate still exists over the meaning of certain parts of the Constitution of the United States.

It can be argued that the *mahele* was inaugurated with the enactment of legislation to establish the Board of Commissioners to Quiet Land Titles on December 10, 1845, although it was not effective until passage of the "Act to Organize the Executive Departments" on February 7, 1846. On February 9, Kauikeaouli appointed the five original members of the board, which was commonly known as the Land Commission: William Richards, who was elected chairman by his colleagues, John Ricord, James Young Kānehoa, John Papa I'i, and Zorobabela Ka'auwai.

Richards had been an official advisor to Kauikeaouli and the chiefs since 1838 responsible for instructing them in the varied facets of Western culture. Since he had been instrumental in convincing Kauikeaouli of the value of private land ownership, his appointment to the Land Commission is not surprising.

When he was asked to assist the Hawaiian leaders in learning about Western culture, Richards had resigned from the American mission in Hawai'i. Nonetheless, he was in effect an integral part of the missionary establishment that was pushing hard for changes in Hawaiian government and life. Without demeaning Richard's integrity or ability, it is worth noting that although he was college educated, he came to Hawai'i with very little experience in government or commerce. He was ordained in 1822 at the age of twenty-five, having graduated from Andover Theological Seminary. That same year, he married Clarissa Lyman and sailed for Hawai'i, where he became one of the pioneer ministers at Lahaina.

There is no question that Richards served Kauikeaouli faithfully and well, but he, like many other influential missionaries, simply did not bring to Hawai'i decades of actual working experience in varied facets of the culture that he was espousing so strongly. In particular, he did not have extensive experience in the professions related to real estate transactions or law. Richards' participation on the Land Commission was rather short, for in 1847 he suffered a stroke and died.

At the time of his appointment, Ricord was the attorney general of the kingdom. A lawyer from Belville, N. J., Ricord had briefly served as secretary of state for the Republic of Texas. He was appointed attorney general in Hawai'i shortly after his arrival in 1844. Having drafted the legislation that organized the government, Ricord was a logical choice for the Land Commission.

Ricord's involvement with the commission and in other positions, however, highlights one of the problems that Kauikeaouli faced in relying on foreigners to help him with the restructuring of the government: The pool of available talent that he could tap was fairly small. He could choose people from the missionary community, many of whom had limited experience beyond their pastoral duties, or he could select businessmen and other professionals who came to Hawai'i for various reasons, many of whom came with less than impressive resumes even if they were exceptionally bright and dedicated people.

Ricord, for instance, was only 32 when he arrived in Hawai'i. Although trained in the law and possessing a well-traveled background, he stepped into a position of great importance without having had the chance to develop a thorough understanding of Hawaiian culture or history. Especially in relation to the events of the *mahele*, it seems doubtful that such a young man with so little experience in Hawai'i could grasp all the nuances of the Hawaiian system of land tenure and then extrapolate to what the impact on the people would be when a system based on European and American legal systems was imposed. Indeed, as young as he was, it is also questionable whether Ricord could have developed a thorough understanding of the American system of land law.

Nonetheless, Ricord's impact on Hawai'i was significant though short-lived. He left the islands in 1847 after disentangling himself from a messy legal issue concerning debt, only three years after his arrival. His spot on the Land Commission was given to Joseph Henry Smith, who had begun his service with the commission as one of its clerks.

Young though Ricord was, he was a wizened professional compared with William Little Lee, who was appointed to the Land Commission in 1847 to replace Richards. In that same year, Lee, a lawyer, was also named chief justice. Born in 1821, Lee was only 26 when he was named to these two important positions. His youth was not offset by an extensive knowledge of Hawai'i, however, for he had arrived in the islands only a year earlier, in 1846. Clearly talented and industrious, he was still very young and inexperienced.

George Morison Robertson was another foreign-born member of the Land Commission, having been appointed in 1850, at age 29, to replace Samuel M. Kamakau. Robertson came to Hawai'i in 1844, but he left for approximately a year when he joined the gold rush of 1849. Although he distinguished himself in service to Hawai'i in many capacities before his death in 1867, he brought little experience in land matters to the commission.

Kānehoa was the son of John Young, an Englishman who had been a trusted advisor of Kamehameha. His brother, Keoni Ana or John Young 2nd, was the *kuhina nui*, or prime minister, of the kingdom at that time. Not only was Kānehoa older than the two foreign lawyers on the Land Commission, he had a long history of involvement in the government, having accompanied Liholiho on the official state visit to England in 1823–24. Like Richards and Ricord,

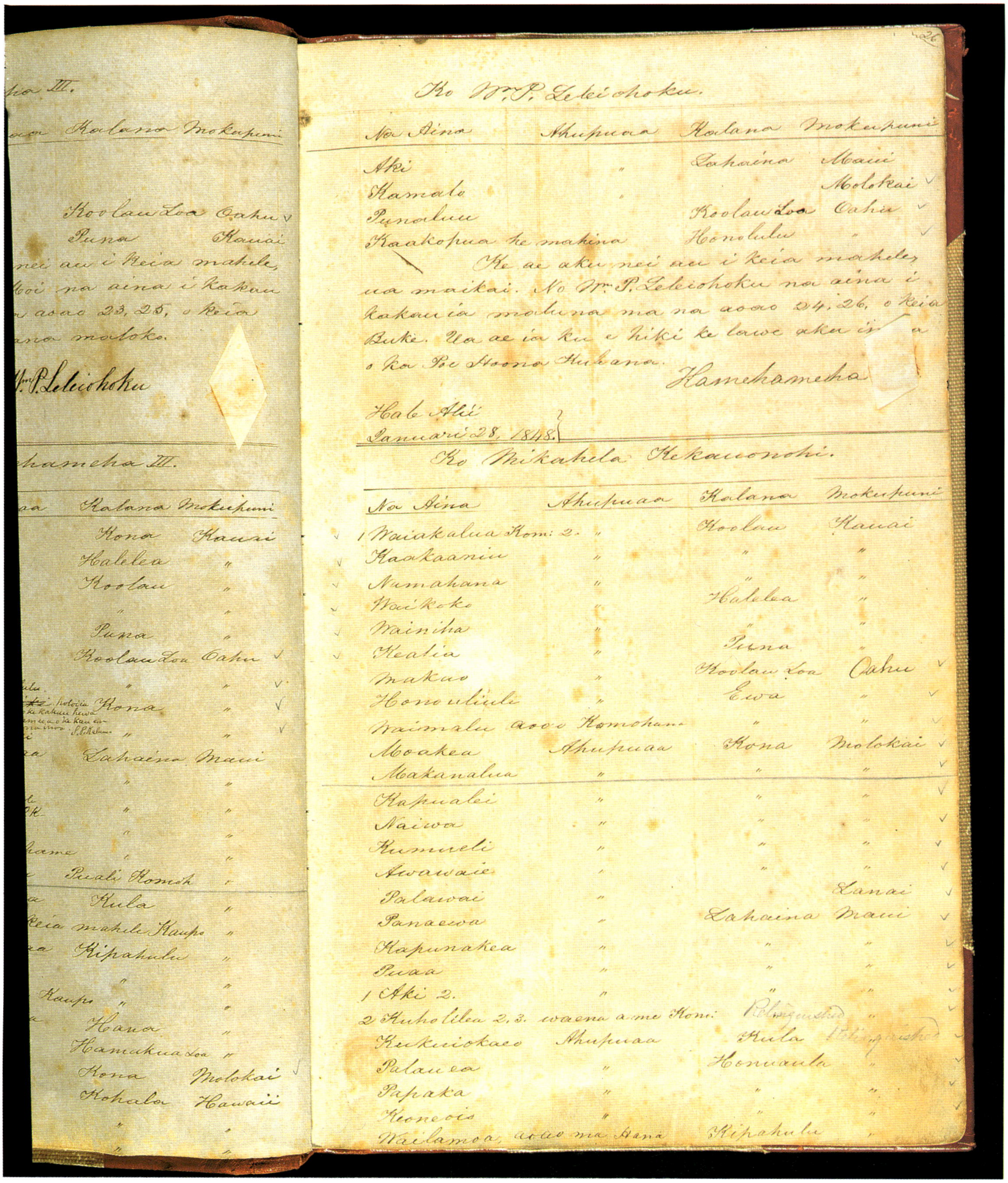

Ko Wm. P. Leleiohoku.

Na Aina	Ahupuaa	Kalana	Mokupuni
Aki	"	Lahaina	Maui
Kamalo			Molokai
Punaluu		Koolauloa	Oahu
Kaakopua ke makina		Honolulu	"

Ke ae aku nei au i keia mahele ua maikai. No Wm. P. Leleiohoku na aina i kakau ia maluna ma na aoao 24, 26, o keia Buke. Ua ae ia ku e hiki ke lawe aku i[...] o ka Poe Hoona Kuleana.

Kamehameha

Hale Alii
Ianuari 28, 1848.

Ko Mikahela Kekauonohi.

Na Aina	Ahupuaa	Kalana	Mokupuni
1 Waiakalua Kom: 2.	"	Koolau	Kauai
Kaakaaniu	"	"	"
Numahana	"	"	"
Waikoko	"	Halelea	"
Wainiha	"	"	"
Kealia	"	Puna	"
Makuo	"	Koolau Loa	Oahu
Honouliuli	"	Ewa	"
Waimalu aoao Komohana		"	"
Moakea	Ahupuaa	Kona	Molokai
Makanalua	"	"	"
Kapualei	"	"	"
Naiwa	"	"	"
Kumueli	"	"	"
Awawaie	"	"	"
Palawai	"		Lanai
Panaewa	"	Lahaina	Maui
Kapunakea	"	"	"
Puaa	"	"	"
1 Aki 2.	"	"	"
2 Kuholilea 2, 3. waena a me Komi		Relinquished	"
Kukuiokaeo	Ahupuaa	Kula	Relinquished
Palauea	"	Honuaula	"
Papaka	"	"	"
Keoneoio	"	"	"
Wailamoa, aoao ma Hana		Kipahulu	

19 Buke Mahele, page 26

Arguably the most important document in Hawaiian history, the *Buke Mahele*, or Mahele Book, is a series of four notebooks in which the division of lands between Kauikeaouli and the *ali'i* of Hawai'i was recorded. Page 26 lists lands award to William Pitt Leleiohoku (1835–1877), brother of Kalakaua and Liliukalani, and also to Mikahela Kekuaonohi (1805?–1851).

From: Board of Commissioners to Quiet Land Titles. *Mahele Book*. 1848.
Source: Hawaii State Archives.

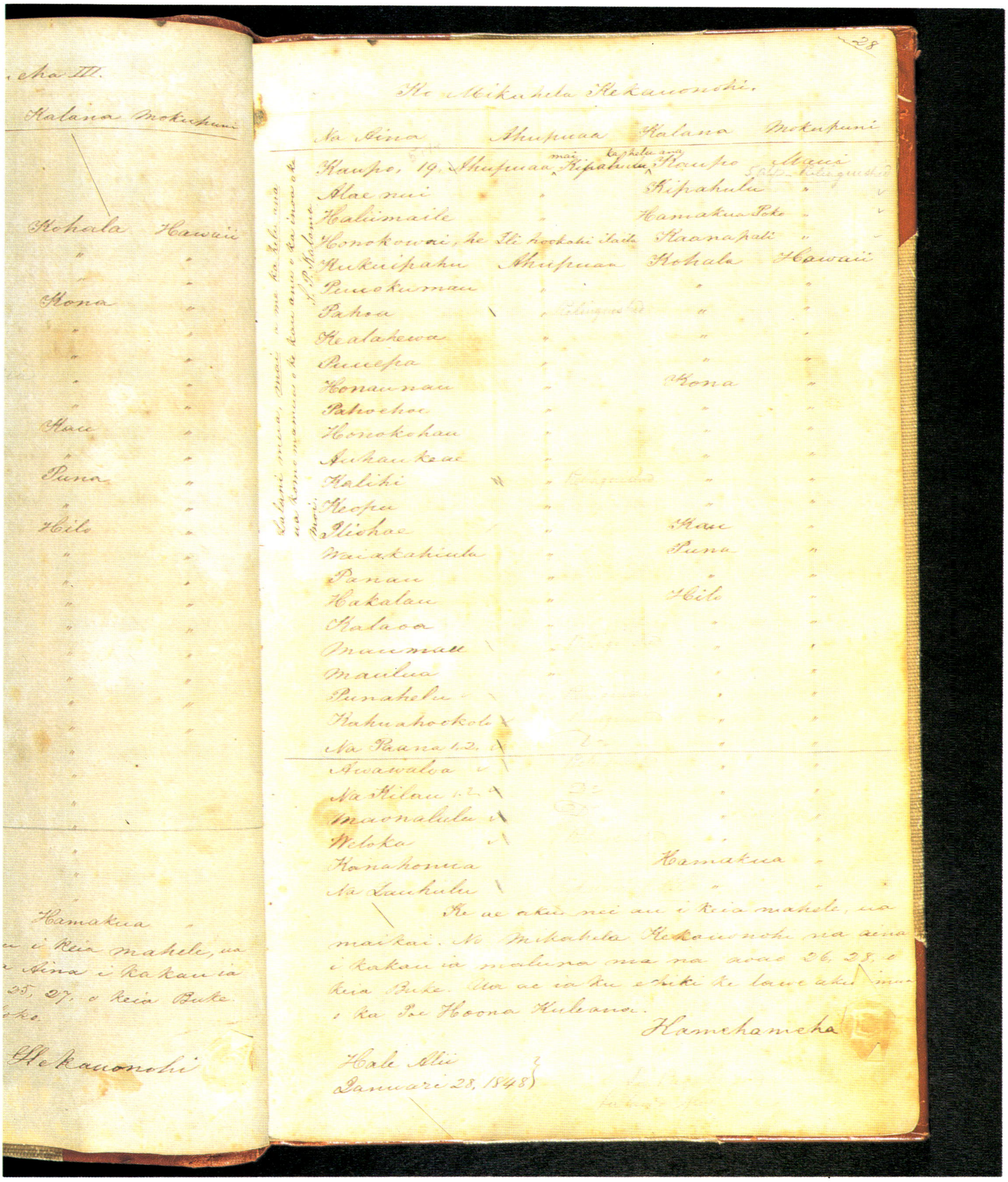

20 Buke Mahele, page 28

This page continues listing lands awarded to Mikahela Kekuaonohi in the *mahele*.
From: Board of Commissioners to Quiet Land Titles. *Mahele Book*. 1848.
Source: Hawaii State Archives.

21 Timothy Ha'alilio (1808–1844) (left) and William Richards (1793–1847)

Richards was elected the first chairman of the Board of Commissioners to Quiet Land Titles, the group of Hawaiians and foreigners who oversaw the technical issues of implementing the *mahele*. Source: Hawaiian Mission Children's Society, Honolulu, Hawai'i; negative number N-1838.

22 William Little Lee (1821–1857)

At the age of twenty-six, Lee was appointed to the Land Commission in 1847, the same year he was named Chief Justice. He arrived in Hawai'i in 1846.

Source: Hawaiian Mission Children's Society, Honolulu, Hawai'i; negative number N-1557.

23 George Morison Robertson (1821–1867)

Robertson came to Hawai'i in 1844 and was appointed to the Land Commission in 1850, when he was twenty-nine. Robertson served Hawai'i in many capacities, including as a judge.

Source: Bernice Pauahi Bishop Museum; negative number CP 102,019.

24 James Young Kānehoa (1797–1857)

Among the original appointees to the Land Commission in 1846, Kānehoa served only until 1847. He was one of the more experienced men in the group, having participated in government affairs since the 1820s.

Source: Bernice Pauahi Bishop Museum, Honolulu, Hawai'i; negative number CP 16972.

Kānehoa served only briefly on the Land Commission before being replaced by Nueku Namaau on August 18, 1847.

John Papa Iʻi was the only member to serve for the entire duration of the Land Commission. The author of a widely respected book, *Fragments of Hawaiian History*, Iʻi also served as associate justice in the Superior Court. Well-educated, Iʻi was in his mid-forties when he was appointed to the Land Commission.

Zorobabela Kaʻauwai served until 1850, when he was replaced by Joshua Kekaulahao, who had served as the Hawaiian secretary to the commission.

Nueku Namauʻu replaced Kānehoa on the commission on August 18, 1847, and died the following year. He was replaced by Samuel Manaiakalani Kamakau.

Kamakau, educated at Lāhaināluna Seminary, is well-known for his voluminous writings on Hawaiian history and for having served in the legislature from 1851 to 1876. Born in 1815, he was 33 when he was named to the Land Commission, having had the benefit of serving briefly as a district judge on the island of Maui.

Certainly the members of the Land Commission were capable and industrious and pursued their work with integrity. Nonetheless, this brief review of their qualifications emphasizes that Kauikeaouli and the Hawaiian people were tackling a task of enormous complexity and scope with little true experience at the helm. As a group, the Hawaiian members seem to have possessed more life experience and more knowledge of the existing land tenure system than did the foreign born members. But they were untrained in the legal systems of the culture from which the new land system was being adapted. On the other hand, the foreign-born members who had some years of life experience under their belt had little formal training in the land practices of their native land let alone their adoptive culture. Moreover, the foreign members with legal training were young and inexperienced.

What was this group of disparate men to accomplish on the Land Commission? An excellent summation of the commission's purpose was given by Chinen in his treatise on the *mahele*:

> The Land Commission had no authority to create any new interest in land; it was granted only the authority to determine the rights in land existing as of the date of the Act. Upon confirmation of a claim and the issuance of an award by the Land Commission to a successful claimant, the statute authorized the minister of interior to issue a Royal Patent on the Award, upon payment of commutation to the government.[4]

Before starting on their task, the members of the commission thoroughly studied the traditional land system and devised a set of principles to guide their decisions, entitled "Principles Adopted by the Board of Commissioners to Quiet Land Titles in Their Adjudication of Claims Presented to Them."

In the preface to the "Principles," the commission stated that a tract of land

> . . . if all parts of it were equally valuable, might be divided into the three equal parts, and an allodial title to one then be given to the lord, and the same title be given to the tenants of one third, and the other third would remain in the hands of the King, as his proportional right.[5]

The meaning of this statement has been debated subsequently. In reality, an *ahupuaʻa* usually did not consist of "equally valuable" land. Only in cases of small *ahupuaʻa* or *ʻili* might the argument be made that all the land was equally valuable. In one sense, the statement seems to imply that the *makaʻāinana* had a right to one-third of the land, but later interpretations generally have not supported that view.

The "Principles" clearly declared, however, that the only persons with vested rights in the land were the *mōʻī*, the *aliʻi* and *konohiki* (who acted as landlords), and the *makaʻāinana*. (A separate legislative act precluded foreigners from gaining fee simple interest in land.) Of the seven specific principles, the first five detailed the process to be followed by the Land Commission. The sixth covered the forms of commutation, or payment, necessary to secure land title. The seventh noted that no land claims made after February 14, 1848, would be acknowledged.

On August 20, 1846, the members of the Land Commission signed the "Principles," which were approved by the legislative council on October 26, 1846. Starting from this date, the Land Commission allowed only eighteen months for the filing of claims. Because claims were not coming in as hoped for and the February 15 deadline was approaching, William Little Lee wrote letters to the missionaries in the outer areas in December 1847 and January 1848 asking them to encourage the Hawaiians to submit claims. He said the Commission would accept almost any form of claim. Submissions then came in by the hundreds.

While it is important to note that claims had to be submitted by the deadline, the Land Commission was not committing itself to rule on those claims within that time period. The gathering of testimony and issuance of awards occurred later.

One thing that held the Land Commission back in its work was the difficulty that Kauikeaouli and the major *aliʻi* had in deciding how to apportion land among themselves, which led to political maneuvering too complex to examine in this volume. By late December 1847, however, Kauikeaouli and the principal *aliʻi* comprising the Privy Council had agreed on guidelines for dividing the lands of the kingdom among themselves. Kauikeaouli

25 John Papa Iʻi (1802?–1870)

Like the other Hawaiians on the Land Commission, Iʻi was middle aged and experienced when he was appointed. He was the only member to serve for the entire term.

Source: Bernice Pauahi Bishop Museum; negative number CP 96424.

would retain all his private lands for himself, thus nullifying the principle of the Constitution of 1840, which said all land belonged to the government. Of what remained, one-third would be assigned to the government, another third to the *aliʻi* and *konohiki*, and the last third set aside for the tenants, or *makaʻāinana*. From this statement, it seems clear that one-third of the land was intended for the *makaʻāinana*. But this is another instance of the "hazy" nature of some of the *mahele* documents, for it is also clear that no effort was ever made to directly assign one-third of the land to *makaʻāinana*. In fact, there was a caveat in the "Principles" that said such land division into thirds "was to take place whenever any chief, *konohiki*, or tenant desired such a division. . . ."[6]

A committee comprised of Keoni Ana, Gerrit P. Judd, Jonah Piʻikoi, and Mataio Kekūanaoʻa was appointed to assist Kauikeaouli and the *aliʻi* in dividing the lands among themselves. From January 27 to March 7, 1848, Kauikeaouli formalized the division of lands among himself and 245 *aliʻi* and *konohiki*. In what has come to be known as the *Buke Mahele* or *Mahele Book*, Kauikeaouli acknowledged the rights of individuals to various land divisions, and they in turn acknowledged his claim to the land he set aside for his own use.

It was this division, or *mahele*, of lands between Kauikeaouli and the higher *aliʻi* which gave the name for the entire process of conversion to the system of private land ownership. A key element in this is that Kauikeaouli's private lands were listed in the same volume as the private lands of the other *aliʻi*. In this manner, Kauikeaouli was emphasizing that such lands belonged to him as an individual and not to the office of the *mōʻī*.

On March 8, 1848, the day after the final transaction of the *mahele*, Kauikeaouli further subdivided lands he had retained for himself. In two statements included in the *Buke Mahele*, he described lands that were to become his personal property and other parcels that he was giving to the government. In the first of these statements, he noted

> Know all men by these presents, that I, Kamehameha III, by the Grace of God, King of these Hawaiian Islands, have given this day of my own free will and have made over and set apart forever to the chiefs and people the larger part of my royal land, for the use and benefit of the Hawaiian Government, therefore by this instrument I hereby retain (or reserve) for myself and for my heirs and successors forever, my lands inscribed at pages 178, 182, 184, 186, 190, 194, 200, 204, 206, 210, 212, 214, 216, 218, 220, 222, of this book, these lands are set apart for me and for my heirs and successors forever, as my own property exclusively.[7]

In the second statement, Kauikeaouli described the properties he was giving to the government as being those listed on pages 179 to 225 of the *Buke Mahele*. Later computations indicate that Kauikeaouli retained 984,000 acres, the government received 1,495,000 acres, and the *aliʻi* were given 1,619,000 acres.

The legislature acknowledged this act on June 7, 1848. Part of the statute passed at that time read

> . . . It hath pleased our Sovereign Lord the King, to place the lands so made over to his chiefs and People, in the keeping of the House of Nobles and Representatives, or such person or persons, as they may from time to time appoint, to be disposed of in such manner as the House of Nobles and Representatives may direct, and as may best promote the prosperity of this kingdom and the dignity of the Hawaiian Crown. . . .[8]

This legislative act, however, noted that such lands were

still subject to the claims of tenants living on them. The most important aspect of the act, however, was that it authorized grant sales of government lands—already authorized by the Act of 1846 establishing the Land Commission.

The *mahele* did not actually convey title to the various *ali'i* and *konohiki*; it essentially gave them the right to claim the lands assigned to them. They were required to present formal claims to the Land Commission and pay a commutation fee, which could be accomplished by surrendering a portion of their land to the government.

Because of the lack of surveyors in the kingdom, on June 19, 1852, the legislature authorized the Land Commission to award lands described in the *Buke Mahele* strictly on the basis of ancient or traditional names. This meant that no survey had to be made but that the integrity of the traditional land units would be honored. It was expected that future surveys would rectify this situation. An interim solution, sponsored by Lot Kamehameha, adopted August 23, 1862, established a commissioner of boundaries to hear testimony on land boundaries conveyed in name only and forbid issuing patents in name only. The actual acreage of these lands, therefore, did not become known until after the establishment of the Hawaiian Government Survey in 1870 to document the boundaries of all the *ahupua'a* and *'ili*. A few of these boundaries have yet to be established.

Not all *ali'i* and *konohiki* understood or were comfortable with the legal aspects of the *mahele*. Consequently, many failed to present their claims to the Land Commission. Through legislative extensions, they were given until June 30, 1862, to present their claims and obtain a patent from the minister of the Interior. Such relief did not resolve the problem, however, and as late as December 16, 1892, the legislature authorized the minister to issue patents to those *ali'i* and *konohiki* who still had not filed their claims. Delinquent claims had to be made by January 1, 1895, or the lands would revert to the government.

Kuleana Lands

While waiting for Kauikeaouli to complete the formal *mahele*, the Land Commission heard nearly seven hundred claims for house lots in Honolulu, Lahaina, and Hilo, and experimented in land sales in Makawao and Mānoa. By mid-1848, the commission was able to embark in earnest on its major task.

As noted earlier, the second principle of the Land Commission agreed to by the Privy Council in 1847 reserved one-third of the lands that Kauikeaouli had not retained for himself for the *maka'āinana*. This principle, however, was never actually executed. In the *Buke Mahele*, the principle is declared that those lands that were set apart as the lands of the Hawaiian government were "subject always to the rights of the tenants." In effect, the principle that the *maka'āinana* were entitled to one-third of the lands of the kingdom was satisfied by two means: the award of *kuleana* grants for land already occupied, and the sale of property outright from the lands set aside for the government.

26 Samuel Manaiakalani Kamakau (1815–1876)

This noted Hawaiian scholar and government official was thirty-three when he was appointed to the Land Commission in 1847. Educated at Lāhaināluna Seminary, he probably had at least a textbook knowledge of surveying and some exposure to the Western system of land ownership.

Source: Bernice Pauahi Bishop Museum, Honolulu, Hawai'i; negative number CP 21360.

Four resolutions approved by the Privy Council on December 21, 1849, and adopted by the legislature on August 6, 1850 gave the *maka'āinana* the right to gain title to lands that they had actively occupied or worked. This in no way amounted to the same as giving them one-third of the lands of the kingdom, but there was a rationale for adopting this course at that time, as explained in a letter written by Robert C. Wyllie, the minister for Foreign Affairs, in which the fear of invasion by a foreign power was discussed

> ... in such a case, it is only private property that is respected, and therefore it would be wise to put every native family throughout the Islands, in possession of a good piece of land, in fee simple, as soon as possible.

If danger comes, we shall then have done our duty in providing for the poor natives, and if it does not come, we shall only do them justice; for by the principle adopted by the Land Commission, the poor natives are entitled to one-third.[9]

The *maka'āinana* were to be awarded fee simple rights to lands that they could prove they had cultivated for at least two years. This land was known as a *kuleana*. Rules prohibited claiming waste land or property farmed merely to increase the size of the claim. The many people abandoning the land to seek economic opportunities in the bustling port towns of Honolulu and Lahaina were thus excluded from land ownership. This inequity was addressed in an editorial written by William Little Lee and published in *The Polynesian* on February 16, 1850, in which he says that Kauikeaouli and the *ali'i* would make land available to those not qualifying for a *kuleana* claim. Although he does not specify the mechanics of how this was to be done, it seems that he was referring to the process of grant sales of government lands that followed.

Regardless of the original idea or principles that were stated, the *kuleana* claim process basically involved four steps: present a claim to the Land Commission between 1846 and 1848; procure a survey of the property; appear before the Land Commission between 1850 to 1855 to testify on behalf of the claim, and submit the award to the minister of the Interior to receive a royal patent. In land records, two identifiers are found for each *kuleana* award—a Land Commission Award (or LCA) number and a Royal Patent (or RP) number.

The claims and descriptions, including the survey plats, are bound into ten folio volumes, which are now in the Hawai'i State Archives.The testimonies submitted to document the awards are in fifty lesser volumes. Along with the *Buke Mahele*, this documentation is the basic record of private property in Hawai'i.

Unlike the *ali'i* and *konohiki*, the *maka'āinana* were not expected to pay a commutation fee to gain title to their rural *kuleana*, yet the survey cost and other fees more than offset that advantage. Theodore Morgan wrote that

> As to the value of land, it is likely that the $6 to $12 fee often exceeded the value of the *kuleana*, even granting the superior fertility of *kuleana* land. Land was exchanging in the late forties and fifties from 25 cents to $1.50 an acre, with $1 about the median.[10]

A variety of sources indicate that two or three dollars went to the surveyor. Since the average size of the *kuleana* claim was about three acres, the cost of the survey was nearly equal to the value of the property.

House lots claimed in Honolulu, Lahaina, and Hilo were treated differently in that *maka'āinana* paid a commutation fee equal to one-fourth the value of the unimproved property.

How effectively this process worked for the *maka'āinana* is still disputed. According to its final report, the Land Commission heard 13,514 claims for land yet made only 9,337 awards. Many of the unawarded claims were duplicates, some were not granted, and about 1,500 were not acted upon, probably because the claimants had died or failed to appear. In 1893, the Hawaiian Government Survey estimated that 28,658 acres of land were awarded as *kuleana*, less than 1 percent of the total area of the kingdom.

Several scholars have pointed out that *kuleana* land tended to be the richest farm land in Hawai'i, whereas the huge acreages given to the *ali'i* and *konohiki* included vast stretches of desert, mountain lands, and forests. Recent attention, however, has focused on the fact that only a small percentage of *maka'āinana* benefitted from the *mahele*. Based upon her extensive research into the *mahele* records, Lilikalā Kame'eleihiwa argues that there were actually 14,195 claims for *kuleana* lands and only 8,421 actual grants. Some were multiple claims by a single person and some were additional claims by *ali'i* and *konohiki*. At the time, she says, there were approximately 88,000 Hawaiians, of whom 29,220 were males over the age of 18. After subtracting *kuleana* awards to *ali'i*, *konohiki*, and foreigners, she estimates that the average *kuleana* award was for approximately three acres.[11] Marion Kelly, in a 1982 Bishop Museum lecture, stated that there were 8,205 awardees, including 200 foreigners, 560 *ali'i*, and 7,500 *maka'āinana*.

A full accounting of the impact on the *maka'āinana* cannot be made, however, unless the number and acreage of grant sales from government lands to native Hawaiians is also taken into account, since it was, as noted earlier, Kauikeouli's intent from the beginning to sell land to *maka'āinana* who were not awarded a *kuleana*. The volume *Index of all Grants and Patents Land Sales* lists the individual grantee, the location, and the acreage for all such transactions. Clearly, the overwhelming majority of the personal names appearing in this list are Hawaiian. According to W. D. Alexander, appointed the first surveyor general of the kingdom in 1870, "between 1850 and 1860 nearly all the desirable Government land was sold, generally to natives."[12] It is also clear, however, that many of the larger sales went to people with non-Hawaiian names. When the quantity of government land sold to *maka'āinana* is included, the impact of the *mahele* may take on a different light than if looked at solely in terms of *kuleana* awards. A total of nearly 400,000 acres in grant sales were recorded between 1846 and 1860, but no analysis of the amount that went to Hawaiians is available.

Theories abound as to why so few claims were submitted for *kuleana* lands. Traditionally, the arguments have suggested that many *maka'āinana* did not understand the legal process, that *ali'i* and *konohiki* discouraged their tenants from applying for the grants, or that the *maka'āinana* simply did not trust the process. Kame'eleihiwa has offered other rationale, including the wish to avoid offending *ali'i* or to maintain traditional relationships.

To date, existing literature has not considered whether people may have failed to complete their claim because they could not obtain or afford the necessary survey. The Land Commission was dissolved on March 31, 1855. Depending on which beginning date one uses, in five to seven years nearly 9,000 pieces of property, or about 1,200 to 1,800 a year, had to be surveyed. Given that there were very few surveyors who had mastered even the crudest knowledge of the profession, this represented a tremendous workload. Although surveys were required for Land Commission hearings from 1850 to 1855, it is possible that claims were accepted without them. Letters from one surveyor, Curtis J. Lyons, imply that his surveys were rarely if ever challenged by the Land Commission members, so little attention actually may have been paid to them.

Impact on the Government

The dissolution of the Land Commission does not mark the end of this important historical period, however. Many of the *ali'i* and *konohiki* did not complete their land claims until near the end of the century. It also can be argued that the period did not close until January 3, 1865, when the legislature officially made crown lands—those lands that belonged to the *mō'ī* personally—inalienable. This did not prevent the government from selling its lands, however.

The act was passed in response to the fact that a *mō'ī* could sell the crown lands or will them to someone else, leaving the successor to the crown with nothing. After the death of Alexander Liholiho in 1863, the potential pitfalls of this situation were recognized. Consequently, his widow, Emma, in exchange for not claiming a portion of the crown lands as inheritance, was given a lifetime pension of 6,000 dollars per annum, and legislation was enacted that prevented the crown lands from being sold or willed to anyone other than the new *mō'ī*. In effect, this act undid Kauikeaouli's efforts to define crown lands as the personal property of the *mō'ī*. What Kauikeaouli had feared and others warned against actually happened in 1893, when the monarchy was overthrown and the crown lands were confiscated by the new government in 1894 at the establishment of the Republic instead of being recognized as the private property of the *mō'ī*.

Land Sales to Foreigners

The legislature granted resident aliens the right to acquire fee simple rights to land on July 10, 1850, some four weeks

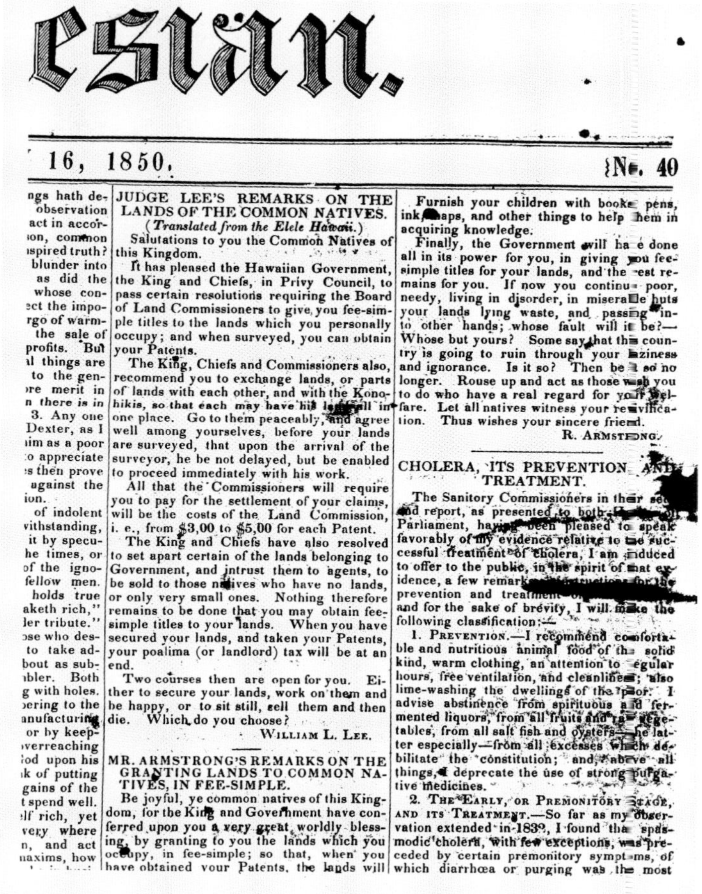

esian.

16, 1850. No. 40

JUDGE LEE'S REMARKS ON THE LANDS OF THE COMMON NATIVES.
(*Translated from the Elele Hawaii.*)

Salutations to you the Common Natives of this Kingdom.

It has pleased the Hawaiian Government, the King and Chiefs, in Privy Council, to pass certain resolutions requiring the Board of Land Commissioners to give you fee-simple titles to the lands which you personally occupy; and when surveyed, you can obtain your Patents.

The King, Chiefs and Commissioners also, recommend you to exchange lands, or parts of lands with each other, and with the Konohikis, so that each may have his [illegible] in one place. Go to them peaceably, and agree well among yourselves, before your lands are surveyed, that upon the arrival of the surveyor, he be not delayed, but be enabled to proceed immediately with his work.

All that the Commissioners will require you to pay for the settlement of your claims, will be the costs of the Land Commission, i. e., from $3,00 to $5,00 for each Patent.

The King and Chiefs have also resolved to set apart certain of the lands belonging to Government, and intrust them to agents, to be sold to those natives who have no lands, or only very small ones. Nothing therefore remains to be done that you may obtain fee-simple titles to your lands. When you have secured your lands, and taken your Patents, your poalima (or landlord) tax will be at an end.

Two courses then are open for you. Either to secure your lands, work on them and be happy, or to sit still, sell them and then die. Which do you choose?

WILLIAM L. LEE.

MR. ARMSTRONG'S REMARKS ON THE GRANTING LANDS TO COMMON NATIVES, IN FEE-SIMPLE.

Be joyful, ye common natives of this Kingdom, for the King and Government have conferred upon you a very great worldly blessing, by granting to you the lands which you occupy, in fee-simple; so that, when you have obtained your Patents, the lands will

Furnish your children with books, pens, ink, maps, and other things to help them in acquiring knowledge.

Finally, the Government will have done all in its power for you, in giving you fee-simple titles for your lands, and the rest remains for you. If now you continue poor, needy, living in disorder, in miserable huts your lands lying waste, and passing into other hands; whose fault will it be?—Whose but yours? Some say that this country is going to ruin through your laziness and ignorance. Is it so? Then be it so no longer. Rouse up and act as those wish you to do who have a real regard for your welfare. Let all natives witness your revivification. Thus wishes your sincere friend.

R. ARMSTRONG.

CHOLERA, ITS PREVENTION AND TREATMENT.

The Sanitory Commissioners in their [illegible] and report, as presented to both [illegible] Parliament, having been pleased to speak favorably of my evidence relative to the successful treatment of cholera, I am induced to offer to the public, in the spirit of that evidence, a few remarks [illegible] for the prevention and treatment [illegible] and for the sake of brevity, I will make the following classification:—

1. PREVENTION.—I recommend comfortable and nutritious animal food of the solid kind, warm clothing, an attention to regular hours, free ventilation, and cleanliness; also lime-washing the dwellings of the poor. I advise abstinence from spirituous and fermented liquors, from all fruits and raw vegetables, from all salt fish and oysters—the latter especially—from all excesses which debilitate the constitution; and, above all things, I deprecate the use of strong purgative medicines.

2. THE EARLY, OR PREMONITORY STAGE, AND ITS TREATMENT.—So far as my observation extended in 1832, I found the spasmodic cholera, with few exceptions, was preceded by certain premonitory symptoms, of which diarrhœa or purging was the most

27 The Polynesian, February 16, 1850

With time running out for Hawaiians to claim their *kuleana* grants, efforts were mounted to encourage the commoners to secure rights to the land they had worked. In addition to the call for quick action, Judge William Little Lee's editorial highlights two important aspects of the *mahele*—that land was to be sold to those with little or no land, and that after claiming their *kuleana*, there would be no more labor tax. The latter provision effectively undermined the foundation of Hawaiian government. Lee's letter reflects the Government's concern that the *mahele* be successful. Portions of a letter from Richard Armstrong, an American missionary, reflect the fervor with which foreigners espoused the value of private land rights.

Source: Hawaiian Mission Children's Society, Honolulu, Hawaii; negative number N-1884.

before it authorized *kuleana* awards to the *maka'āinana*. The naturalization of foreigners had long been a sensitive subject among the Hawaiian people, and this act basically ended the debate over how foreigners should be treated in regard to land. As a result of these various actions, much of the land of Hawai'i found its way into the control of non-Hawaiians between 1850 and 1865 through purchase or lease.

Such land sales were often vast, encompassing entire *ahupua'a* or *'ili*. In many cases, a survey was made to accompany the legal transaction, several of which are reproduced in Chapter 5. Although these transactions do not technically fit into the concept of the *mahele*, they are nonetheless part of the privatization of land in Hawai'i.

28 William Patterson Alexander (1805–1884)

Possibly responsible for the introduction of land surveying in Hawai'i, Alexander taught the subject to his students at Lāhaināluna Seminary on Maui. He also surveyed for Hawaiians during the *mahele*. His son became the first and only Surveyor General of the Hawaiian Government Survey.

Source: Hawaiian Mission Children's Society, Honolulu, Hawaii; negative number N-1156.

4 The Surveyors of Hawai'i

Surveying was probably introduced to Hawai'i in January 1779, when Captain Cook's expedition spent several weeks at Kealakekua Bay on the island of Hawai'i. The detailed inset of the bay that appeared on the map published in the official narrative of the expedition (EMH, Ill. 2) was certainly compiled using surveying techniques. Other early European explorers applied their surveying skills to the production of other detailed charts of the islands. Notable among these are the chart of Honolulu harbor compiled under the direction of Otto von Kotzebue in 1817 (EMH, Ill. 25) and that of southern O'ahu drawn by Lt. Charles Malden in 1825 (EMH, Ill. 33).

Kotzebue's narrative contains a vivid account of how the use of flags to mark prominent harbor features created a stir that almost led to a violent confrontation. Malden's work produced no such furor, but his efforts notably included the earliest known use of a long baseline—two points between which the distance was measured very precisely—to provide an accurate framework to which all features he mapped could be referenced. Regrettably, the existence or location of Malden's baseline was forgotten, for there is no indication that surveyors during the *mahele* took advantage of it as a way to reference their own surveys. In the latter part of the 19th century, the Hawaiian Government Survey devoted extensive resources to the establishment of baselines to improve the surveying in the islands. Had surveyors during the *mahele* period used the baseline Malden created, the surveying work in the complex and economically significant areas of Waikīkī and Honolulu would have been better and more consistent.

The surveying practiced by the maritime explorers was not the same as land surveying, however. Hydrographic surveying is concerned with charting coastlines and hazards to navigation. Land surveying, as noted earlier, is concerned with the description of the extent of real estate.

Missionaries and Surveying

Surveying had long been an important occupation in the country from which the Hawaiian missionaries came. Notable historical figures such as George Washington, Thomas Jefferson, James Madison, and Henry David Thoreau were accomplished surveyors. The missionaries who came from the United States brought the skill with them. Some of them conducted surveys out of general interest. As noted in *The Early Mapping of Hawai'i,* the maps drawn by Ursula Emerson, which have been dated to 1833, were obviously compiled from unknown surveys. The large map of Hawai'i compiled at Lahainaluna Seminary in 1838 also must have benefitted from surveys made in the islands, since it is much more accurate than the existing published maps from the voyages of Cook and Vancouver.

Lahainaluna Seminary can be considered the focal point for geographical studies in Hawai'i in the era of the *mahele.* Established in 1831 on a hill overlooking the Maui town of Lahaina, the seminary was a high school for Hawaiian youth and young men. Lorrin Andrews, the founder and first instructor, placed great emphasis on geography in the curriculum there. To support his teaching goals, he oversaw the translation of standard geographies into the Hawaiian language. In addition, he introduced the practice of copperplate engraving, partly to support the production of maps for his geographies.

Another missionary, William Patterson Alexander, joined Andrews at Lahainaluna in 1843. Alexander is important to the surveying of the *mahele* in two ways. He taught surveying to a number of his pupils, or scholars, as they were referred to then, and he actually did a substantial amount of surveying in support of *kuleana* claims himself. Alexander figures in the history of surveying in Hawai'i in two other ways—his eldest son, William DeWitt Alexander, became the first surveyor general of Hawai'i when the Hawaiian Government Survey was established in 1870, and his grandson, Arthur C. Alexander, was a noted surveyor in the islands.

William Patterson Alexander was born in Paris, Kentucky, in 1805. He attended Centre College in Kentucky and Princeton Theological Seminary in New Jersey. On October 12, 1831, he was ordained in Cincinnati, Ohio. On October 25 of that same year, he married Mary Ann McKinney, and the next month set sail from Bedford, Massachusetts, for Hawai'i. He was briefly assigned to the station at Hilo, then went for a year to the Marquesas Islands in an unsuccessful attempt to establish a mission there. On their return to Hawai'i in 1834, the Alexanders were stationed at the Wai'oli mission station in Hanalei Valley on Kaua'i. In 1843, he was assigned to work at Lahainaluna Seminary.

Alexander instructed a number of young Hawaiians in the art of surveying and during the *mahele* spent Saturdays surveying *kuleana* in the area around Lahaina. Part of the rationale for this work was clearly monetary. With seven children to support, and the oldest nearing college age, Alexander needed more money than he was receiving as a missionary. The overseeing authority for the Hawaiian missionaries, the American Board of Commissioners for Foreign Missions, frowned on this effort as being inappropriate for a man of the gospel and communicated its displeasure in a letter in 1850. Alexander responded with a strong letter that set forth several reasons why he should be allowed to continue his surveying work.

> In your later letter, you take the position that surveying the land is incompatible with the missionary relation. Now I have employed most of my Saturdays during the past year in surveying, & I have received compensation for it. I think I have not injured the missionary influence by this work, but by it have conferred a kindness on the people, which they have received as a kindness, have improved my health & thus rendered myself more valuable as a missionary, & obtained something to aid in the education of my children.
>
> Indeed I think it doubtful whether I could have held on to my post during the year without the active labor which surveying has led me to perform.
>
> Unless convinced that I am doing wrong, I expect, as I have opportunity, to continue the same course in time to come.
>
> I have written plainly and frankly hoping thus to secure a clear understanding.[1]

The clear understanding was not achieved, apparently, for in 1850 Alexander took a year's leave of absence from his missionary position to improve his health and devote full time to his surveying. Most of his surveying was conducted in three areas: around Lahaina, in the Hāna area on the eastern end of Maui, and in the cool *kula*, or highlands, on the slope of Haleakalā.

Another missionary who applied his skills as a surveyor was John Emerson, assigned to the Waialua mission station on the north shore of O'ahu. As noted elsewhere, his wife Ursula compiled three maps that were based on original surveys, and Joseph may well have been a contributor to those maps.

John Emerson and William Alexander shared one thing in common. They each fathered a son who became a prominent surveyor in Hawai'i. As recounted in *The Early Mapping of Hawai'i*, a Hawaiian once compared the elder Emerson to his five sons with the following comment:

> Then there is Joe, a surveyor for the Hawaiian government, but your father not only surveyed government lands, he sold them to the people and helped them get homesteads.[2]

In the first stage of the *mahele* in 1848, when the *maka'āinana* were given the opportunity to submit claims for the land they worked, Emerson helped many of the Hawaiians in the Waialua area prepare their documents. He noted that

> I must show them how to do it; they do not know how to do it themselves, and there is none but myself to do it for them &, if not done before the 14th of Feb it can not be done at all.[3]

Emerson actually went beyond the role of surveyor and became the land agent for the government in the Waialua district. He chose a different course than most land agents, for he actively sought Hawaiians to buy property that he surveyed. According to his own accounts, he surveyed approximately 8,000 acres into some 300 lots ranging in size from 50 to 1,000 acres.

Emerson also shared Alexander's streak of independence. In 1850, he wrote a letter to the same American Board of Commissioners for Foreign Missions that had questioned Alexander's surveying activities in which he placed the work of surveying and land sales into perspective. He basically dismissed criticism of his actions rather succinctly:

Perhaps some will say I have forsaken my calling to become land agent for the Government in the sale of lands to the natives. Be it so.[4]

Emerson gave three reasons for his work as land agent:

> First, the land was for sale and if some one who loved the people did not take up the business, it would probably be sold to their disadvantage in great lots to a few purchasers who would become lords of the land, and such a result would soon render my stay in Waialua almost useless.[5]

His second reason was more mundane: He simply thought that it would not detract from his missionary duties. His third reason was more profound:

> A third reason that led me to accept this appointment is that the people asked to have it assigned to me, and my acceptance has secured to a large number of poor people lands that they could not have obtained otherwise, and a saving of cost for services which they could not have met.[6]

Emerson stated that his whole compensation amounted to a tract of land of 15 acres valued at $5 each. While that may have been true in 1850, between 1851 and 1867 he purchased or otherwise gained rights to 2,500 acres, for which he paid $4,699 dollars. As with Alexander, however, it should be noted that the relationship of the missionaries to the American Board of Commissioners to Foreign Missions had become very strained at the time of life when he had to think about ways to provide for his family's future. Like Alexander, Emerson eventually resigned from the mission and supported himself through various means.

Alexander and Emerson may be exceptions when their surveying accomplishments are compared with those of other missionaries. Nonetheless, missionaries such as Edward Bailey, Dwight Baldwin, Artemas Bishop, Jonathan Green, Daniel Dole, George Berkeley Rowell, and Asa Thurston are known to have engaged in surveying work, either during or after their official relationship with the Board of Commissioners.

29 John S. Emerson (1800–1867)

Along with his wife Ursula, he administered the mission station at Waialua on the island of *O'ahu*. Emerson applied his rudimentary knowledge of surveying to help the Hawaiians during the *mahele*. He also functioned as the government Land Agent in that part of the island.

Source: Hawaiian Mission Children's Society, Honolulu, Hawaii; negative number N-1738.

Hawaiian Surveyors

It is known that a number of Hawaiians practiced surveying during the *mahele*. Regrettably, not nearly as much biographical information has been published for these men as for the missionaries and sons of missionaries who engaged in the profession. According to Arthur C. Alexander's comments, the proficiency of the group of Hawaiians seems to parallel that of the foreigners, running the gamut from very good to mediocre.

Until more research is published regarding these Hawaiians, Alexander's remark that all of the Hawaiians learned surveying from his grandfather, William Patterson Alexander, at the Lahainaluna Seminary should be accepted with some skepticism, as there is reason to suspect that some of them may have been trained by Lorrin Andrews, a man possessing an exceptional range of talents who was there before Alexander arrived in 1843.

Two Hawaiians are worth mentioning as surveyors because of their professional reputation and the important roles they played in the Hawaiian government, Samuel P. Kalama and John W. Makalena. Kalama, who generally signed his surveys as S. P. Kalama or merely S.P.K., appears to have been a remarkable individual. It is not known when Kalama was born, which leads to some speculation about his attendance at Lahainaluna Seminary. In the obituaries published in *The Friend* and *The Advertiser* after his death on December 2, 1875, it was remarked that he was about 60

30 Surveying instruments of John S. Emerson

Some of the tools that Emerson used are preserved in the collections of the Hawaiian Mission Children's Society. Shown are stakes and the surveyor's chain. The loops at each end are handles by which the chain can be pulled taut.

Source: Hawaiian Mission Children's Society, Honolulu, Hawai'i; photograph by David Franzen.

years old. If he was truly close to that age, then he likely would have been at Lahainaluna before William Patterson Alexander arrived there. Although Kalama is a prominent name, there is reason to suspect that he may have been the student at Lahainaluna who engraved a number of maps that were published there, including the important large map of the kingdom issued in 1838. It carries the authorship statement, *"Na Kalama i kaha,"* which could be interpreted as either "drawn by" or "engraved by" Kalama.

Lending further credence to the possible identification of S. P. Kalama as a young man with great interest in geography is a reference in his obituary in *The Friend* to his saving Dr. Gerrit P. Judd's life at Kīlauea in 1841. In that incident, described on page 94 of *The Early Mapping of Hawai'i*, Dr. Judd was attempting to scoop lava from Kīlauea with a frying pan when a sudden rise in the flow nearly engulfed him. According to the official narrative of the expedition by Charles Wilkes, he was pulled to safety by an unnamed Hawaiian. Knowing that Kalama accompanied Judd to the island of Hawai'i with the U. S. Exploring Expedition, which was actively engaged in surveying and mapmaking, suggests that he already possessed some skill in those areas, which in turn reinforces the notion that he may have been the one responsible for the maps drawn at Lahainaluna under Andrews.

In addition to doing a significant amount of surveying in support of the *mahele*, Kalama later held important posts in the government. He served several terms in the Legislature and was a member of the Privy Council under *mo'i* Lot Kamehameha, Lunalilo, and Kalākaua. Not mentioned in his obituary is a position he held of profound importance to the *mahele*; he was the clerk who recorded the various entries in the *Buke Mahele*. This important task involved more than mere transcription. There was a complicated phase during which Kauikeaouli and the *ali'i* were negotiating which lands belonged to whom. There are a number of lists found in scattered files in the Hawai'i State Archives in which Kalama recorded the lands being kept and given up by the various people involved. In many cases, the lists show that he was responsible for making additions and lining out other names as the negotiations progressed. The two pages from the *Buke Mahele* that are illustrated are in Kalama's hand.

Contemporary surveyors hold Kalama's work in as high esteem as did Arthur C. Alexander in 1920. One of his surveys, that of the *ahupua'a* of Waikāne, is reproduced in the following chapter.

The other Hawaiian surveyor noted here, John Makalena, also served in several government capacities. After his death in early December 1877, his obituary in *The Advertiser* reported that he had served as commissioner of Rights of Way and Water, a Circuit Court judge, and on the Privy Council. In his work as a surveyor, he served the four kings spanning from Kauikeaouli to Kalākaua. One of Makalena's most important surveying works was a map of downtown Honolulu on which he delineated the lands belonging to various *ali'i*, which is reproduced in the following chapter.

Foreign-Born Surveyors

Many of the surveyors who worked during the *mahele* were men who had come to Hawai'i for a variety of reasons. It is evident from the work they did that a number of them were neither educated nor well-trained as surveyors. Three foreign-born men, however, stand out as surveyors of exceptional skill.

William C. Webster, born in 1827, was possibly the most competent surveyor and civil engineer in Hawai'i at that time. As a young man in his native Scotland, he had apprenticed for three years to a land surveyor, and thus became highly qualified in this field. Upon his arrival in Hawai'i in 1850, he was appointed by a fellow Scotsman, Robert Chrichton Wyllie, Minister of Foreign Affairs, to the position of land agent for Kauikeaouli's own lands. In that capacity, he produced a large number of surveys throughout the islands, several of which are reproduced in Chapter 5.

Webster quickly became fluent in Hawaiian and was reportedly quite sympathetic to the Hawaiian people. In 1855, he was elected to the House of Representatives, in which he served until his death in 1864. In September 1862, he was elected speaker of the House. In addition, he was appointed to the Privy Council and acted as the secretary to Alexander Liholiho, Kamehameha IV.

Another foreigner with impressive skills was Rudolf Wilhelm Meyer. Born in Hanse, Germany, in 1826, Meyer studied hydraulics and surveying in the civil engineering curriculum at the high school in Hamburg. He worked for a brief period with the city's water works before coming to Hawai'i at the age of 24 in 1850. He ended up on Moloka'i, where he met a young *ali'i* woman named Kalama Waha, whom he married in March, 1851.

Meyer became the government surveyor for Moloka'i and did considerable surveying work in that capacity. In addition, he filled a number of government functions on that island, including superintendent of the leper colony at Kalaupapa. Meyer also managed the Moloka'i Ranch for a succession of owners, including Alexander Liholiho, Princess Ruth Ke'elikōlani, Princess Bernice Pauahi Bishop, and Charles Reed Bishop. An example of Meyer's surveying is reproduced in Chapter 5.

Theophilus Metcalf, a native of either Geneva or Canandaigua, New York, was a Harvard-educated civil engineer and a trained surveyor before he arrived in Hawai'i on May 19, 1842. At age 21 he had been chief engineer of the Michigan Central Railroad. Due to ill health, he was advised to move to a warmer climate. He married Kaililikapuno of Kukāniloko, O'ahu, and established the

31 Rudolph Wilhelm Meyer (1826–1897) ca. 1890

Born in Germany and trained as a civil engineer, Meyer was one of the few men in the islands with a solid surveying background. He practiced surveying primarily on Moloka'i, where he settled.

Source: Hawai'i State Archives, negative number 20,603.

first daguerreotype studio in Honolulu in 1845. That business failed in 1846, just as the Land Commission was created. During the next several years, Metcalf performed a considerable number of surveys on various islands. He also served as marshal of the Hawaiian Islands in 1849 and 1850. From 1852 to 1855, Metcalf served as superintendent of public works for the Kingdom of Hawai'i. In 1855, he established a sugar plantation and lumber mill on 1,309 acres he had purchased in 1852 at Kaupakueo, now known as Pepe'ekeo, near Hilo, Hawai'i. Metcalf died in Oakland, California, on a trip to improve his health on August 6, 1866.

Missionary Children as Surveyors

The missionary contributions to the surveying of the *mahele* extend beyond their own work—their children also made significant contributions to the surveying of the *mahele*. In particular, Curtis Jere Lyons and Henry Munson Lyman should be noted. Not only was their work remarkable, but their comments about their experiences contribute important insights into the political and social issues that surround the *mahele*.

These two missionary children were almost that, children. Lyons, for instance, was only 17 when he was hired out of Punahou, the missionary school on O'ahu, by the Land Commission to conduct surveys for the *mahele*. Lyman, Lyons' schoolmate, was a mere 16 when he became a government surveyor with responsibility for the rugged southern half of the island of Hawai'i.

Lyons' first survey, conducted on November 28, 1850, for a man named Kaiwi who lived near the current intersection of Judd and Liliha Streets at the mouth of Nu'uanu Valley, is reproduced here. For a number of weeks, Lyons worked the Honolulu and Waikīkī areas, finally earning enough money, at $2 per *kuleana*, to buy his own surveyors' compass. With that experience under his belt and his own equipment in hand, Lyons was assigned to the island of Hawai'i.

In 1889 or 1890, he prepared a paper describing his experiences while engaged in surveying. Sections of his comments are quite revealing about the nature of the work and shed tremendous insight on some of the problems inherent in converting from the traditional land tenure system to one based on private ownership. During his first duties on Hawai'i, young Lyons worked in the largest valley on the island, which essentially marks the divide between the Hāmākua and Kohala *moku*. He wrote:

> Waipio Valley, with its crowded native population and its infinity of taro patches held me three long, laborious months. The chiefs' men fought the kuleana system as hard as they could. Disputes, at first referred to Mr. Robertson, afterwards came back to me, with orders to do the best I could and act on my own responsibility. A good lesson it was.[7]

The disputes to which he refers were probably arguments between the *konohiki*, or overseer, and the *maka'āinana* who were claiming the *kuleana* they worked. Since the lands awarded to *ali'i* in the *mahele* were subject to the claims of *maka'āinana*, many *ali'i* or their *konohiki* tried to keep *maka'āinana* from applying for their *kuleana* or contested the substance of the claim that was made. In the case of Lyons, he was clearly surveying *kuleana* that had been claimed, and he found himself in the middle of many disputes. The Mr. Robertson he refers to was a member of the Land Commission. Lyons, then 18 at the most, found himself being asked to resolve differences of opinion between the *maka'āinana* and the *ali'i* or *konohiki*:

> Well, responsibility came again—how much land to give the natives in cases where they planted now here, now there, so as to rest the soil from year to year.

32 Curtis Jere Lyons (1833–1914) ca. 1850

Only seventeen and a student at Punahou on O'ahu, Lyons was recruited in 1850 to survey property for the Land Commission. He did most of his work in the Hamakua district on the island of Hawai'i.

Source: Bernice Pauahi Bishop Museum, Honolulu, Hawai'i; negative number CP 112,930.

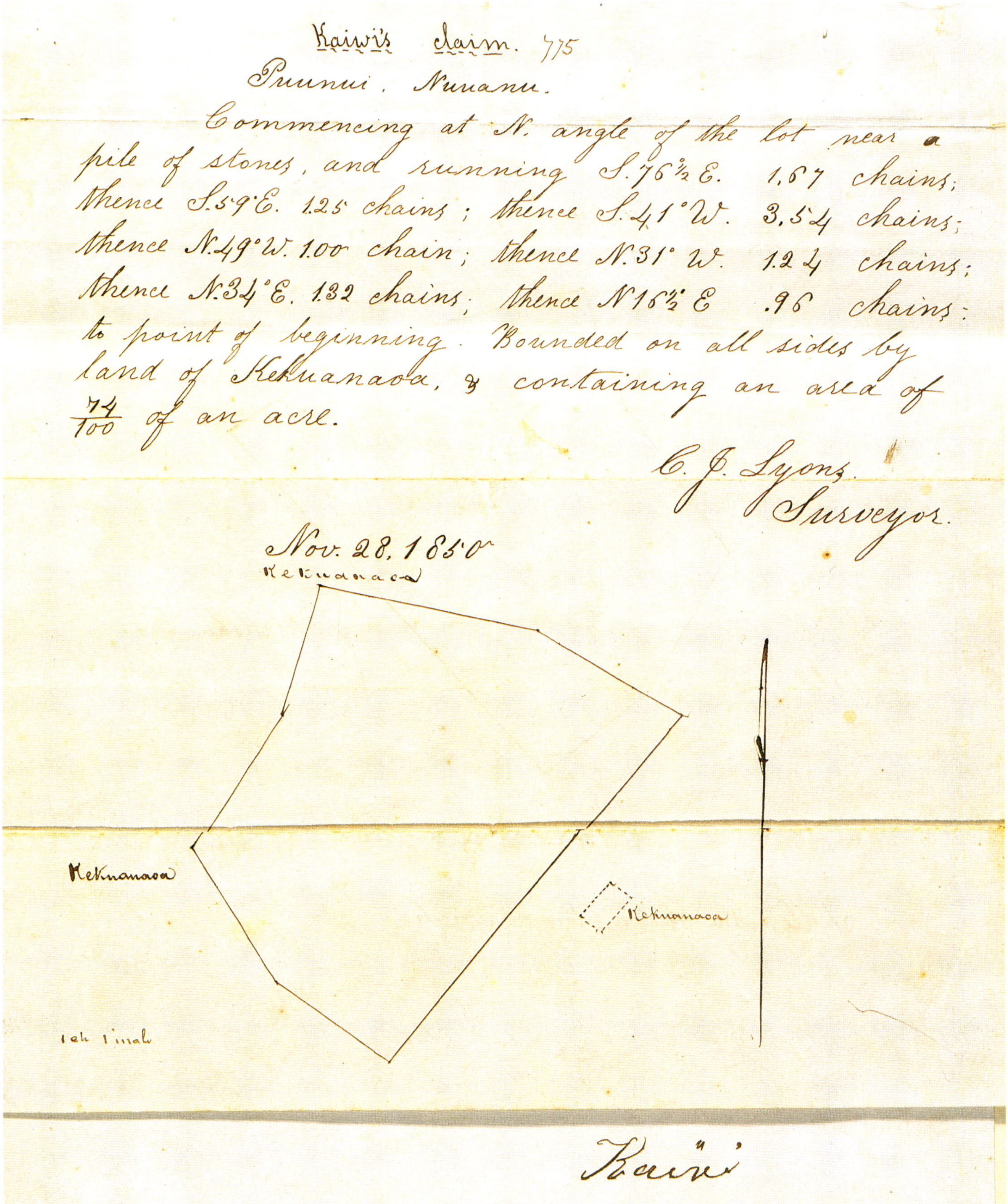

33 Survey of Kaiwi's land by Curtis Jere Lyons, November 28, 1850

Lyons worked on O'ahu for a few weeks before heading to the island of Hawai'i on behalf of the Land Commission. This plat represents the very first survey by the young son of missionaries. The land is near the intersection of present-day Judd and Liliha streets in Honolulu.

25 x 20 cm.; scale 1:792 (1 chain = 1 inch).
Source: Hawai'i State Archives, Royal Patent 213.

> Some agents gave only the particular patches then under cultivation—stingy they—while others, too generous, gave from twenty to forty acres. The rule was to give what the natives improved. I multiplied the average cultivation by the number of years of fallow, or rest—three acres by three—and decided upon ten acres as a fair number, so the Hāmākua natives have an average of ten acres to the kuleana. It was of no use to wait for orders from headquarters, so I acted for myself, and was approved.[8]

So Lyons found himself in the situation of having to interpret the Hawaiian land tenure system to ascertain what was fair. Surveying, therefore, was not merely a matter of measuring angles and lines, it involved listening to people about their use of land and making judgments that would affect the quality of their lives.

For young men like Lyons, the intense and extensive experience of working with Hawaiians certainly contributed to their knowledge of the country they called home and the people who lived there. Surveying forced relationships on these young men that they would not likely have had.

> Hamakua, the present rich cane country, next claimed me. The compass was a magic mystery to the natives, who followed all day long; but the chain, the "kaulahao" was to their minds the authoritative instrument. To their eyes, it marked the line on the ground as it dragged along. The flags reminded them of the ancient annual procession of Ke Akua Makahiki, and they compared the tripod as it folded up, to the bundled bones of their ancestors.[9]

Elsewhere, Lyons notes that

> The leading natives of Hamakua were always generous hosts, which was greatly appreciated.[10]

He adds,

> Sometimes for six weeks or more at a time, I would not have occasion for one word of English, save in writing.[11]

Lyons describes how he induced his schoolmate, Lyman, to give surveying a try:

> Henry Lyman, after more study at Punahou, decided to follow suit. With Munson Coan as a companion, he footed it over the palis of Hilo to Waimea, and back to Hamakua. We all footed it in those days. We practiced together for a few months, visited the top of Mauna Kea, sent Coan home, traversed under the waterfalls to the far valley of Honopue, beyond Waimanu; and then Lyman went to work in his own wild, watery, gulch-divided Hilo.[12]

For Lyman, getting started as a surveyor was exceptionally easy. When William Little Lee of the Land Commission came to the island of Hawai'i, he was so impressed by the young man who had helped Lyons that he offered him two positions: that of government surveyor and land agent for the Hilo, Puna, and Ka'ū *moku* almost on the spot. Lee gave Lyman a letter authorizing him to receive the necessary surveying implements from the Honolulu office of the Interior Department. Upon reaching Honolulu, Lyman discovered that the minister of Interior, Keoni Ana, was far more concerned about the youth's age than was Lee. Keoni Ana insisted on waiting until Lee arrived back in Honolulu to confirm the appointment. Because of Lyman's age, a compromise was reached: Lyman's father was appointed the land agent and allowed to appoint his son as the surveyor for the district.

In his autobiography, Lyman describes the nature of his initial work with Lyons:

> Breakfast and prayers were finished by sunrise; and a little later we started afield wading through the dew-laden grass, to begin work where it was left off the evening before. The claimant of the land under survey acted as guide, and usually carried a long bamboo pole, surmounted by a white flag, which he planted successively at the corners of his estate, while the surveyor took sights with his compass and directed along the boundaries our course with the measuring-chain. Sometimes the track was obstructed by thickets and jungles of rank vegetation, rendering it necessary to send a couple of swordsmen in advance of our column. This pioneering work was entrusted to a pair of stalwart natives, each armed with a cutlass and an axe with which they made short work of all obstacles. Our task was laborious, but not monotonous; for there were numerous intermissions during which we lay at rest on the turf, while our chief took testimony regarding disputed boundaries and made decisions between rival claimants.[13]

For Lyons and Lyman, their work was not the equivalent of a summer job, something done to earn a little money and quickly forgotten. After describing his travails at surveying, Lyons noted at one point that "three years went on this way." Lyman adds some color to his description of how busy he was:

> Orders for surveying came in as fast as they could be executed, and I was continually at work. Dressed in a blue woolen sack-coat with dark checked figure that

34 The Lyman brothers (from left): Frederick Schwartz (1837–1918), Henry Munson (1835–1904), and David Brainerd (1840–1914)

Henry was only sixteen when he was recruited as a surveyor by his friend and classmate, Curtis Jere Lyons. Although his father had the actual title to preserve appearances, Henry acted as Land Agent for the districts of Hilo, Puna, and Ka'u, an area almost the size of Delaware.

Source: Hawaiian Mission Children's Society, Honolulu, Hawai'i; negative number N-1837.

> was almost a plaid,—warm in rain, and cool in the sun,—an unstarched shirt that displayed a wide rolling collar and a flourishing neck-tie, with a bright red sash of Chinese silk around my waist, and a pair of high boots outside of my trousers, the whole outfit surmounted by a soft gray felt hat of the latest Californian pattern, and followed by two faithful Hawaiian chain-men [assistants who measured the distances with the surveyors chain], I strode up and down the country from the northern boundary of Hilo to the farthest cape of Kau, and became one of the most widely known personages in the land.[14]

For both young men, the work came to an end at the same time. Lyons states that

> At the end of 1853, both Henry Lyman and I laid by the compass and chains to rust unused, while we entered upon a higher education, beginning with the usual voyage around Cape Horn.[15]

Lyons, then 20, and Lyman, only 18, completed their careers as surveyors and sailed on the same ship to begin studies at Williams College in Massachusetts.

The commentary of men like Lyons and Lyman is very important to understanding some of the underlying tensions involved in the *mahele* . Whereas much scholarly and popular discussion about the *mahele* hinges on the fact that it was the vehicle through which foreigners ultimately gained control of much of the land of Hawai'i, there were some genuine disagreements between the *maka'āinana* and the *ali'i* who controlled the land. Comments from letters that Lyons sent to George Robertson refer to some of these difficulties. In a letter of March 12, 1851, he discusses two separate issues.

> There is a hau grove on the land claimed by one of the natives, & which is claimed as having been planted by his father (or uncle). The konohiki at first claimed it as his, because, as he said, some one of the people told him that it was a *koele* grove. The people all denying that it had ever been worked by the koele labor

until his day,—he then proposed to divide it. The claimant would not consent. The konohiki then proposed to cut it down & when it grew up again it should return to them. This was refused as he had already had several cuttings & so the matter rests.

I have made some inquiry among the older natives around & those from adjoining places, what they say all unites to prove that the claimant's father did actually plant the disputed hau grove.[16]

In another letter, dated February 18 but with no year given, Lyons again alludes to the expectations that *maka-'āinana* held:

> Accompanying this are 27 surveys; with one which you received before 28. You will see that 4 of them are large claims. Palea, Kanehailua, Auwae, and Paukeaho, and I hear that the natives refer to these as instances of what they ought to have. Palea & Auwae's claims I surveyed according to instructions. Paukeaho has a large piece of 50 acres enclosed by a fence on one side & a deep ravine on the other. I took in no farther than the fence extended. Kahehailua's has a wall on 2 sides but still takes in a good deal of waste land & may have to be diminished a good deal yet although it is now a small part of what he originally claimed.[17]

In the letter quoted above, dated March 12, 1851, Lyons touches on another issue that was obviously of concern to the *maka'āinana*—what became of the effort they were expending on their land as the *mahele* was taking shape:

> If a man has walled in his yard since I have been here in the valley but before I come to survey it / i.e. if the survey is not purposely delayed; has he not a right to what is walled in.[18]

Lyons also notes in this same letter an issue that suggests that many Hawaiians had difficulty with the process of filing a claim:

> what [should be done] when a man has more patches under cultivation than are written down, as is sometimes the case.[19]

The surveyors in the field, such as the young Curtis Lyons and his colleagues, were forced to deal with circumstances that the men who conceived the *mahele* probably never dreamed of. Out of necessity they made decisions on the spot which affected the quantity of land that many Hawaiians received. It is probably fair to suggest that men such as Lyons and Lyman, who grew up in Hawai'i, made it their lifelong home, and were raised with an ethic to support, uplift and protect the Hawaiians, applied the best of their judgment in good conscience.

It is unlikely that Lyons was the only surveyor to encounter the difficult situations that he described. A real question that arises, therefore, is whether surveyors who were new to the islands, those who possessed neither a command of the Hawaiian language nor Hawaiian customs, even recognized such conflicts let alone possessed the knowledge to resolve them justly.

Arthur C. Alexander's List of Surveyors of the Mahele

There is no definitive list of all the men who surveyed property in Hawai'i during the *mahele* . The most extensive listing that has been published was compiled by Arthur C. Alexander, who was the manager of American Factors' Land and Survey Department. The list, which included commentary about each surveyor, was appended to a paper read before the Honolulu Social Science Association on March 1, 1920, entitled "Land Titles and Surveys in Hawaii."

Given that Arthur Alexander's list was compiled some 70 years after the *mahele,* his comments regarding individual surveyors do not come from first hand-acquaintance with them. In some respects, however, that makes his commentary more valuable because a surveyor's reputation rests largely on how well his work is received in subsequent generations. Surveyors may have little occasion to interact with their contemporaries, whereas they may find themselves constantly having to correlate their work to what was done by those who preceded them. In that light, Alexander's comments may represent professional evaluation of a surveyor's work rather than evaluation of the surveyor as a person. Nonetheless, the evaluations that follow still represent the opinion of only one person.

Of the 33 surveyors listed, only 10 were Hawaiians. Eleven men in the list were missionaries or sons of missionaries. According to Alexander, "most, if not all of the early native surveyors were trained at Lahainaluna School under W. P. Alexander." Of these men in general, Alexander noted that "while not always reliable, they were never guilty in their kuleana surveys of such grossly inaccurate work as was done by some of the white men. They also had a great advantage over many of the white surveyors in their intimate acquaintance with Hawaiian land matters and the language."

Alexander, W. P. One of the most careful surveyors of that time.

Bailey, Edward. Work was fairly good; main fault was the correcting of errors of closure in the office without testing on the ground.

Baldwin, Dwight. Surveyed only one or two small pieces in Lahaina.

Bishop, Artemas. Had no conception of the value of accuracy or the desirability of making adjoining surveys agree, consequently his surveys are extremely inaccurate and inconsistent.

Dillon, James. Work was fairly good; used an engineer's theodolite and the magnetic north of the initial point, a method which has the fault that a blunder in reading or recording the magnetic bearing of the first course may swing the whole survey through an angle of several degrees.

Dole, Daniel. Surveyed only a few small kuleanas in Waikiki.

Emerson, John S. The accuracy of his work was impaired by the employment of an unreliable chainman, who, in staking out land sales, Joseph S. Emerson reports, was in the habit of placing the pin in the ground beyond the end of the chain, thus giving more land than the calculated area called for.

Fuller, John T. An extremely careful surveyor; both Joseph S.Emerson and E. D. Baldwin, who have had much experience in rerunning his surveys say that he was the most accurate surveyor of his time.

Gower, John T. A very careless surveyor.

Hopu, Asa. This surveyor evidently used a compass that was quite "off center," as his surveys have to be swung about 4 degrees counterclockwise to fit the ground.

Kahema, Job. Work was poor.

Kalama, S. P. One of the most reliable native surveyors of that time, with a very extensive knowledge of the names and boundaries of Hawaiian lands.

Kalanikahua, D. As far as I can learn, not a very reliable surveyor.

Kaona, J. Surveyed only a few small kuleanas near Honolulu.

Keohokalole, Abraham. Surveyed only a few small kuleanas in Wailuku, Maui; work was revised by Edward Bailey.

Kittredge, Chas. S. A well-trained surveyor; work was not as good as one would expect from his training.

Lyman, Fred S. A very careful surveyor.

Lyman, Henry M. Like his brother, a very careful surveyor; said by C. J. Lyons to have used the average magnetic north in writing out his descriptions.

Lyons, Curtis J. Perhaps the most careful and conscientious surveyor of that time; used the "average needle" in his descriptions.

Makalena, John W. Work fairly good, except when he attempted to survey large tracts.

Meyer, R.W. Said to have been educated in Germany as a civil engineer; a very careful and intelligent surveyor.

Metcalf, Theophilus. One of the good surveyors of that time; described by C. J. Lyons as "a very shrewd and practical man, whose surveys have the merit of always exhibiting and referring to natural features for fixing the lines run." His compass is said by the same authority to have read about 50' to the east of magnetic north, so that his surveys should be corrected this amount before being run out.

Nahale. Did some surveying in Wailuku, Maui; work revised by Edward Bailey.

Pease, W. H. One of the most careless and unreliable surveyors of that time.

Pelham, John. Another very unreliable surveyor.

Polapola, John. Only made a few surveys of small pieces; work said to have been fair.

Richardson, George. Work said to have had the same fault as his brother John's (see below); they may have used the same compass.

Richardson, John. Must have used a very defective compass; his distances are good, while his bearings in most cases are quite unreliable.

Rowell, G. B. Only did a limited amount of surveying at Waimea, Kauai.

Thurston, Asa G. Work was fair.

Turner, A. F. Said to have used an English theodolite, and like James Dillon and Wm.Webster, to have written his notes out in terms of magnetic north of the initial point. His surveys as a rule fit together and close well, but are not easy to rerun, many of them bearing strong evidence of having been "doctored."

Ua, L. S. Work was good for a native surveyor; like Kalama and Makalena, Ua had an intimate knowledge of Hawaiian lands and boundaries.

Webster, William. Perhaps the best trained and qualified civil engineer in the islands at that time; a very careful surveyor, using a theodolite and the initial magnetic north.

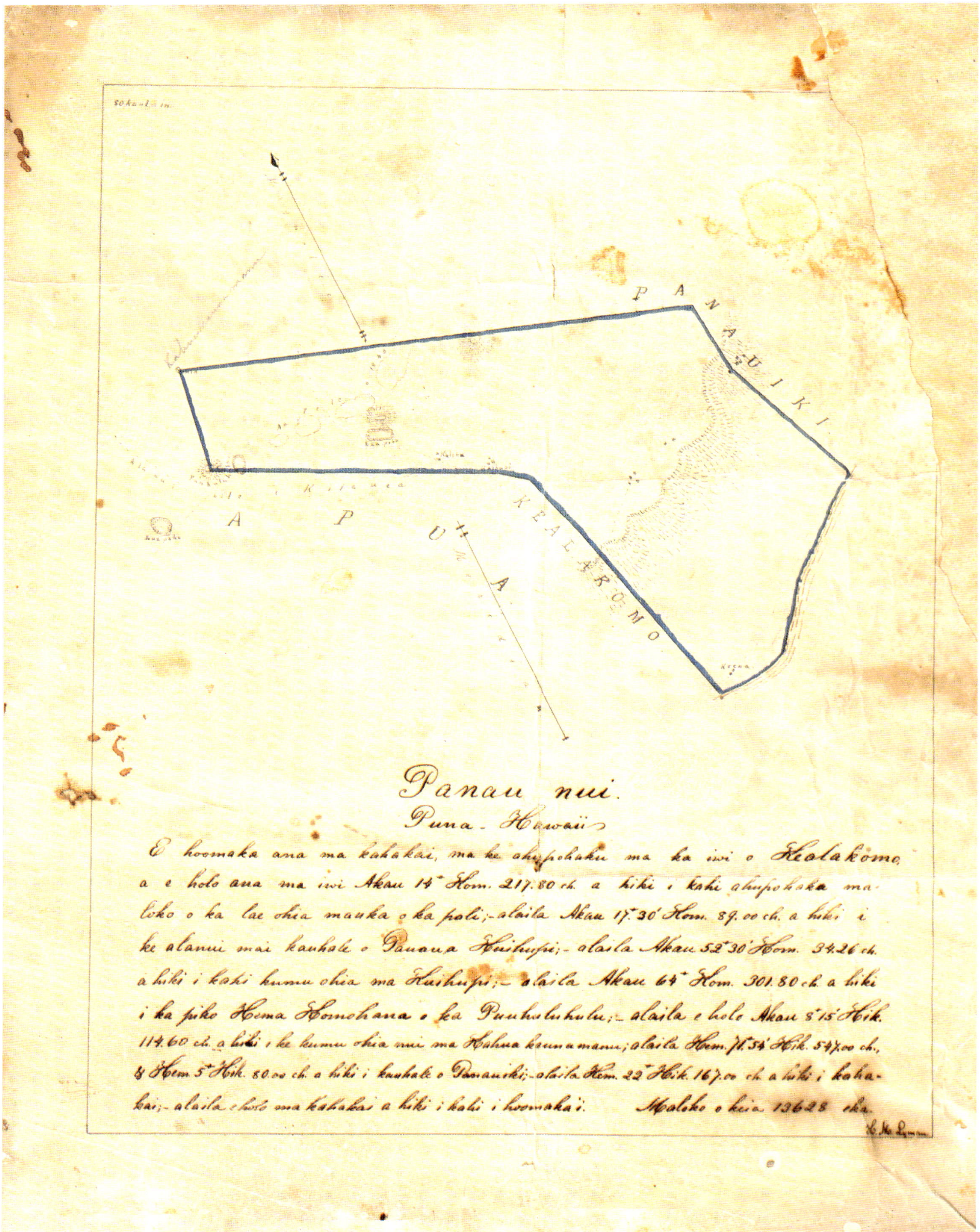

35 Pānau Nui

Surveyed by Henry Munson Lyman. 43 x 35 cm.; scale 1:63,360 (80 kaul to one inch).
Source: Hawaii State Survey Division; reg. map no. 33

5 Surveys from the Mahele

THE HEART of *Surveying the Mahele* is the collection of representative examples of *mahele*-era surveys and plats from sixteen locales throughout Hawai'i presented in this chapter. Where appropriate, historic or current photographs and maps have been used to help convey a sense of the situation in which the surveying was done. Some examples were chosen because they represent typical situations encountered by *mahele*-era surveyors. Others were selected because they are areas familiar to many readers, and thereby embody a relatively common frame of reference. At least one area of each of the inhabited islands is portrayed. The largest of the following sections relates to the *'ili* of Wailele in Mānoa Valley in Honolulu. There is a rich body of surveys of this parcel of land that illustrates many technical and legal aspects of surveying. Each section of this chapter is treated independently, with the text being designed more as extended captions than a continous narrative.

Pānau Nui

Pānau Nui is an *ahupua'a* of some 13,628 acres in the Puna District of the island of Hawai'i. The entire *ahupua'a* now lies within Hawai'i Volcanoes National Park. In fact, a considerable portion of Pānau Nui has been covered by recent lava flows, destroying all villages named on Lyman's plat. Distinguishing characteristics include a significant petroglyph site located at Pu'uloa near the coast, and the Chain of Craters Road, which traverses Pānau Nui as it descends the Hōlei Pali and then runs along the coastal portion of the *ahupua'a*.

In the *mahele*, Pānau Nui, described as Land Commission Award 11216, Apana 41, was awarded to Mikahela Kekau'ōnohi, granddaughter of Kamehameha, a wife of Liholiho, then of Aarona Keali'iahonui, and finally the wife of Levi Ha'alelea. In the *Buke Mahele* she received sixty-seven lands, but she surrendered seventeen to the government in lieu of commutation. Upon Kekau'ōnohi's death in 1851, her lands were inherited by her husband, Levi Ha'alelea. The *Indices of Awards* indicate that there were no *kuleana* claims awarded within Pānau Nui. Indeed, within the whole Puna District there were only four claims of property for a combined total of less than one thousand acres.

Lyman indicated one coastal settlement of three buildings within Pānau Nui, identified as Kaena. Near the crater that Lyman calls Luapono, which is probably Makaopuhi Crater, are two settlements, Kalehu and Kaihui. He also indicates a trail above the cliff called Hōlei Pali, along which a few structures are shown.

Lyman's plat features a description of the property in Hawaiian and indicates that bearings are given to magnetic north. He does not, however, indicate what the variation of magnetic north is from true north. Within this region, it is quite likely that there was considerable variation due to the influence of volcanic activity beneath the surface. A decade earlier, the U. S. Exploring Expedition mapped this same area (EMH, Ill. 56). The resulting map, while very detailed, contained many errors in the geographic relationship of topographic features. Given that this was clearly a difficult area to survey and that Lyman was no more than nineteen years old at the time, this map may well contain some errors.

John Wesley Coulter's maps indicate that about one hundred fifty people lived in this area at the time of the 1853 census. Archeological investigations by the National Park Service suggest that the settlements may have been seasonal. By 1912, when the first U.S. Geological Survey maps of the area were made, no habitations were shown.

36 'Opihilala

Surveyed by C. J. Lyons, April 26, 1852. 58 x 46 cms.; scale 1:9,600 (800 feet to one inch).
Source: Hawaii State Survey Division; reg. map no. 18.

‘Opihilala

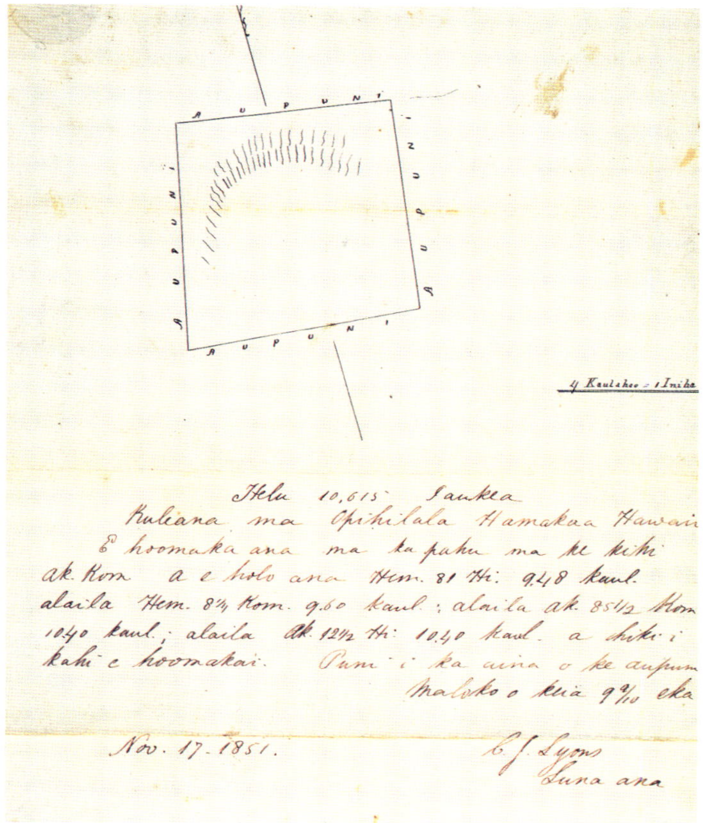

37 [Kuleana of Iaukea]

Surveyed by C. J. Lyons, Nov. 17, 1851. map 31 x 20 cm.; scale 1:3,168 (4 kaul to one inch).
Source: Hawaii State Archives; Royal Patent 6197.

This survey represents a case of government land that was sold to a non-Hawaiian citizen of the Kingdom. ‘Opihilala is an *ahupua‘a* along the Hāmākua coast of the island of Hawai‘i, bounded by Hauola toward Hilo and by Manienie toward Honoka‘a. In the *mahele*, it was designated as government land and subsequently sold to C. J. Lyons, the government surveyor who made this survey and others along the Hāmākua coast from 1850 to 1853. The transaction is identified in government records as Grant 947.

Current maps show that ‘Opihilala measures approximately one-half mile at its greatest width, but only three-tenths of a mile along the two hundred foot high cliff, described by Lyons as “precipitous coast,” which marks its seaward boundary. It narrows to a point at its *mauka* end, approximately three miles inland. Its western edge is bounded by a gulch approximately sixty feet deep. *Makai* of the highway the stream is named Waipunahina Gulch, and *mauka* of the highway it is named ‘Opihilala Gulch. The land rises gently from the ocean and reaches an elevation of about two thousand feet at its furthest extent inland. This property is easily identified when one is driving through the area—the town of Pa‘auilo ends in its western extent at the boundary of ‘Opihilala. The “government road” identified on Lyons’ survey, which is the main street in the town, still runs across ‘Opihilala and makes the curve into the gulch as it did in 1852. Virtually the entire property has been in sugar cane production until recently.

Lyons’ title clearly indicates that this represents a part of the *ahupua‘a* of ‘Opihilala. In fact, the property he purchased encompassed almost the entire land division, apparently ending where the slope became rather steep, as indicated by the series of hachures at the *mauka* end of his survey. According to Lyons’ notes, the land was open with scattered clumps of trees. He refers to “Patches of Sumach or Neneleau,” near the “government road” and scattered growth of trees at the *mauka* end of the *ahupua‘a*. He also refers to a stand of *hala* trees in the neighboring *ahupua‘a* of Hauola, possibly so that anyone using the plat could locate the approximate boundary line along this long, straight section.

Lyons’ survey shows the boundary of the only *kuleana* claim that was made within Opihilala, a 9.9 acre tract awarded to a person named Iaukea. It is interesting to note that no house is indicated on the *kuleana*, but three houses are indicated elsewhere in the *ahupua‘a*.

A significant feature of this plat is the series of straight lines marking the boundary of the property through the ‘Opihilala Gulch. In the Hawaiian concept of land divisions, it was unusual for streams to be used as dividing lines. Rather than showing all of the meanderings of the stream, however, Lyons followed conventional surveying practice by defining it as a series of straight lines. Along this boundary, Lyons notes the location of two springs, Waipunahinu and Alaohia. Such annotations on survey plats provide sources of place names no longer found on standard topographic maps.

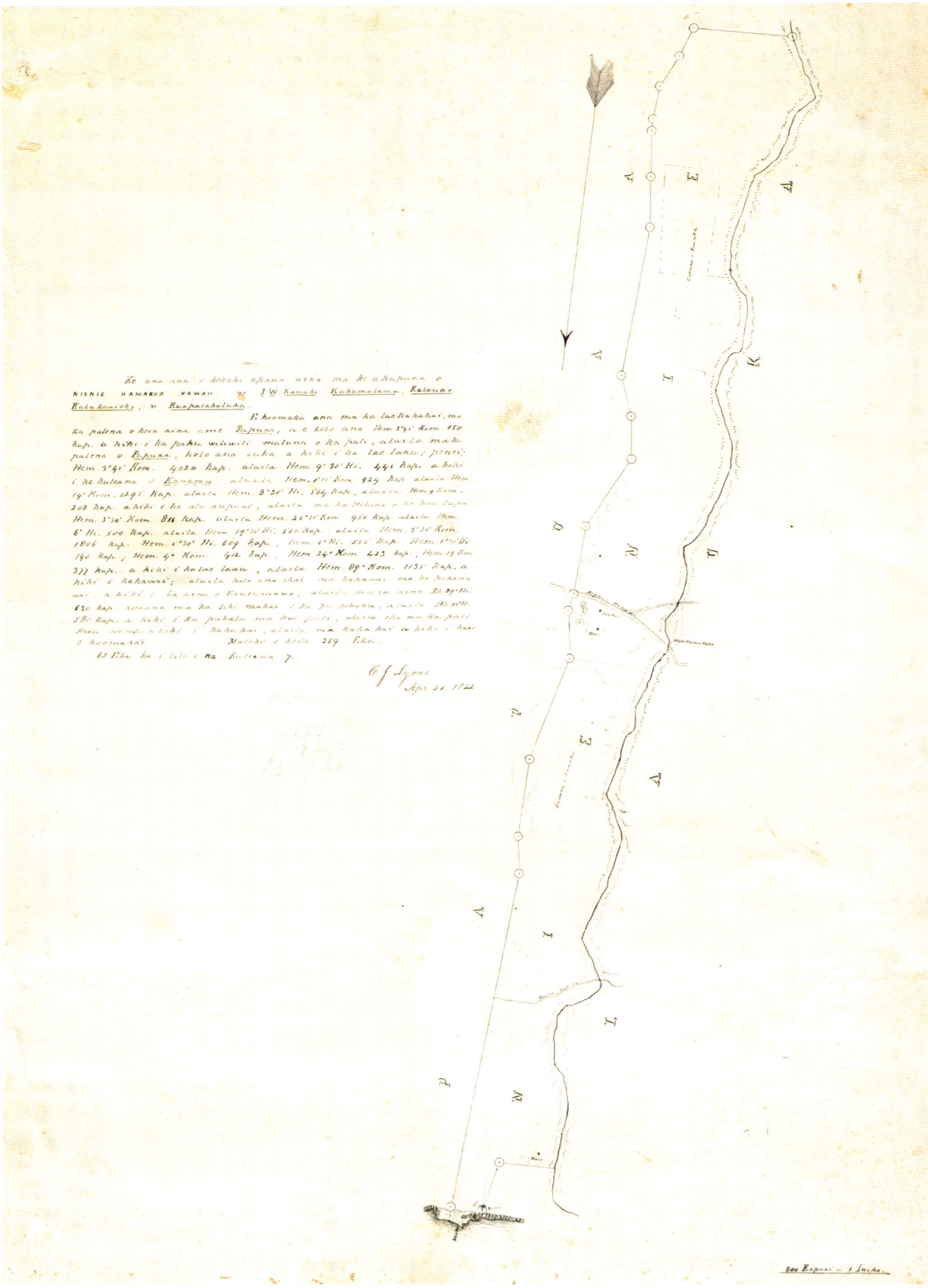

38 Ni'eni'e

Surveyed by C. J. Lyons, April 26, 1852. 50 x 35 cm.; scale 1:9,600 (800 feet to the inch).
Source: Hawaii State Survey Division; reg. map no. 24.

Ni'eni'e

Ni'eni'e—meaning "sheer and smooth, as a cliff"—is a narrow *ahupua'a* on the Hāmākua Coast of the island of Hawai'i near the town of Honoka'a. It is bounded on the Hilo side by the *ahupua'a* of Pāpua'a and on the Waipi'o side by Lauka. The *ahupua'a* stretches inland from the ocean about three miles and is about one-quarter mile wide for its entire extent. Cliffs nearly one hundred feet high mark part of the coastal extent. Today, the site below Honoka'a known as Honoka'a Landing is found on the western boundary of the coastal strip of Ni'en'ie.

The western boundary of Ni'eni'e is formed by a gulch of the same name. The eastern boundary is marked by a series of straight lines which essentially divide in half the land between Ni'eni'e Gulch and the stream just to the east, which is unnamed on present day topographic maps.

Designated as government land in the *mahele*, Ni'eni'e was sold to a *hui*, or group, of Hawaiians, a transaction officially identified as Grant 949. On this survey by Lyons, purchasers are identified as J. W. Kauahi, Kuhumalama, Kalauao, Kuapalahalaha, and Kalauaioho. Lyons' survey records the size of the parcel as 359 acres, although slightly more than seventy acres had been awarded as *kuleana* to eight individuals. This plat appears to show only seven *kuleana*, two clearly indicated as being "kuleana o kanaka." Land records show that these eight *kuleana* were awarded as follows: Hooulu, .43 acres; Kane, 20 acres; Kaulumano, 8 acres; Kea, 9 acres; Kiniakua, 14.1 acres; Polohili, 9.8 acres; Tito, .68 acres; and Unahiolea, 8.9 acres.

Unlike the plat of 'Opihilala illustrated previously, Lyons' description of the land is written in Hawaiian. In the surveys that have survived, the description is generally in the language of the purchaser.

Lyons' survey indicates that not all of the *ahupua'a* was sold to the group of Hawaiians. Excluded was a small section along the coastline now occupied by Honoka'a Landing. While there was no town or industry in existence in 1852 that would have made such a landing site economically important, it is possible that the government recognized that the paucity of harbors or landing sites along the Hāmākua Coast made this location significant and thus excluded this small portion of land from the sale.

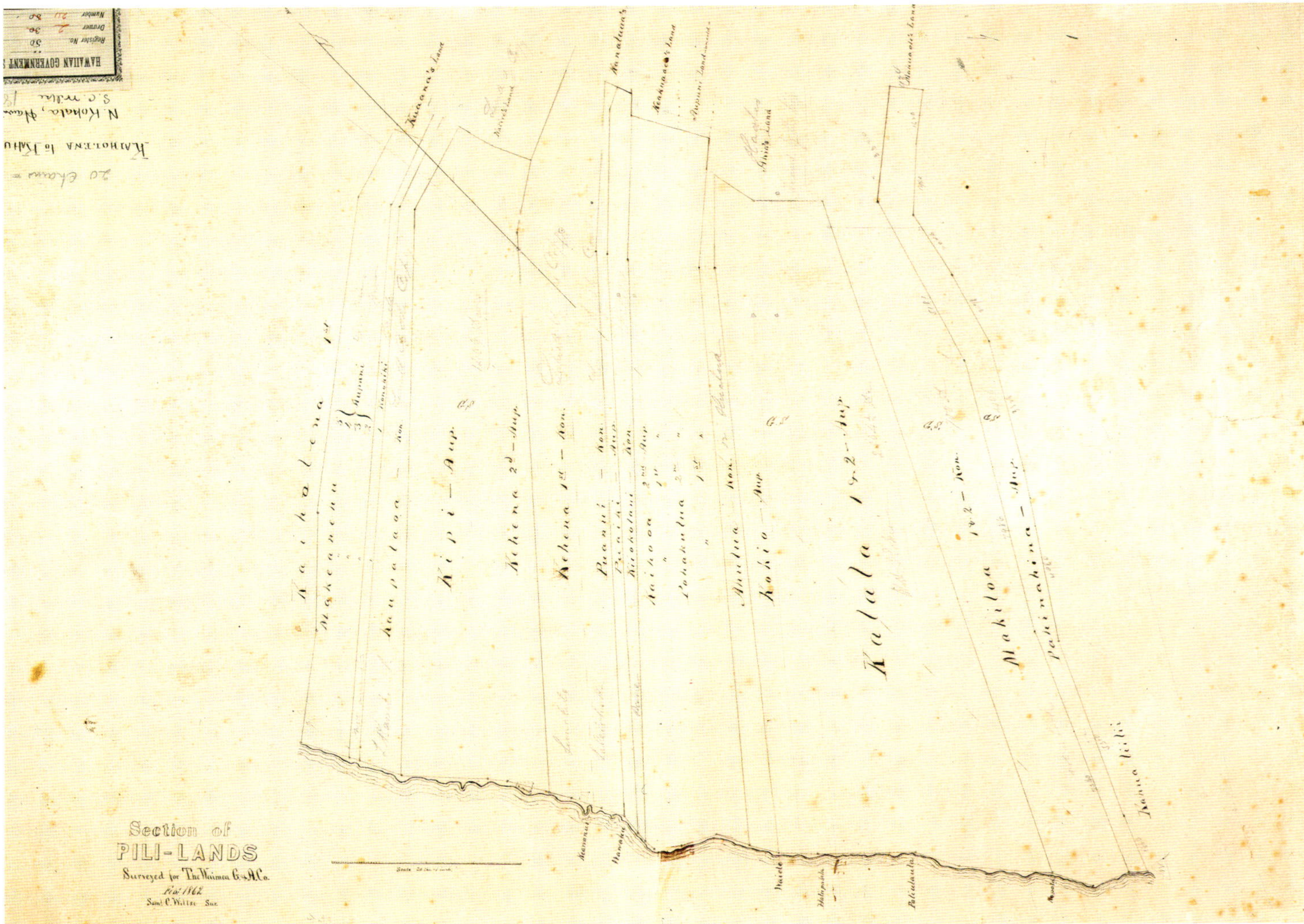

39 Section of Pili-Lands Surveyed for the Waimea G. & A Company

Survey by Samuel C. Wiltse, February 1862. 54 x 72 cm.; scale 1:15,840 (20 chains to the inch).
Source: Hawaii State Survey Division; reg. map no. 50.

Section of Pili-Lands Surveyed for the Waimea G & A Company

Pili is a grass that was used for thatching by native Hawaiians. The arid conditions on the leeward coast of Kohala in the northern part of the island of Hawai'i were conducive to the growth of *pili* grass.

The area encompassed by the survey, extending from Kaiholena in the north to Kahu'a in the south, is remarkable for the number of long, narrow *ahupua'a*. Pahinahina, for instance, is barely one-quarter mile wide and almost six miles in length. A few others are between one-quarter and one-half mile wide. The widest, Kaiholena and Kalala, are about one and one-quarter mile wide. The general uniformity of the local geography provides little indication of how the boundaries were determined. The gulches and other landforms are not well-pronounced. The most distinctive aspect of geography is elevation; above the two thousand foot contour vegetation tends to be greener than below. Recently uncovered archeological evidence from the Lapakahi State Park indicates that small communities existed along the arid coast but tended gardens in the greener upland, or *kula*, areas. Coastal sites were chosen for habitation be-

40 Old Hawaiian ruins, ca, 1925

Southwest Kohala, between Mahukona and Kawaihae, looking east.
Source: Bernice Pauahi Bishop Museum, Honolulu, Hawai'i; negative number CP 112,931.

cause of access to fishing and freshwater sources.

This lifestyle appears to have quickly disappeared after European contact. In the *mahele* few *kuleana* were claimed in this area. Wiltse indicates only one *kuleana* in Kehena, and the U.S.G.S. survey of 1911–1913 shows only two dwellings along this coast. The *Index to Grants* shows that grants were issued in all the *ahupua'a* that were assigned to the government in the *mahele*. Within the last hundred years, the inhabitants of these lands have generally lived in the *kula* portions. The Reverend Asa Thurston, the missionary assigned to Kailua on the island of Hawai'i, reported that in 1823 there were more people living on inland plantations than on the seashore in this area.

Wiltse's survey is very general, designed to indicate boundaries of the lands, whether they were government or *konohiki* lands, and in which lands the Waimea Grazing & Agriculture Company had an interest. Because so many different lands are shown and there is no written description of the boundaries, it is unlikely that this survey was made for any particular grant. The title implies that it was made for the Waimea G & A Company, but it somehow became a part of the records of the Hawaiian Government Survey.

Makawao

Makawao is the name of a land-bound *ahupua'a* on the northwest slope of Haleakalā on Maui. Metcalf's survey encompasses only the area near what is now the town of Makawao, which lies at an elevation of approximately sixteen hundred feet and is among the best agricultural land in the islands. The highly productive area, blessed by good soil, moderate temperatures, and reliable rainfall, today boasts numerous garden farms providing quality produce for Hawaiian markets.

The lands portrayed in this survey were among the first parcels in the islands to be sold to individuals. As described in Chapter 3, the early activities of the Land Commission were limited pending the final distribution of lands to the *ali'i* by Kauikeaouli. During that period, in 1846, the Land Commission experimented with the sale of property in Mānoa Valley on O'ahu and the lands shown here at Makawao.

The Makawao sale was initiated during a tour of the island by Kauikeaouli in December 1845 and January 1846. The entire district, with the exception of the plantation owned by McLane—shown by Metcalf to comprise two parcels totaling 688.94 acres—was to be sold in fee simple. The missionary Jonathon Smith Green, pastor at Makawao, acted as the agent for the sales. Of the Makawao project, Green wrote in *The Polynesian* on September 26, 1846,

> The district is offered for sale as an experiment on the people—as an incitement to awaken industry and thrift—as a test to the question—'Can the common people of Hawai'i be induced to abandon their listless, improvident habits, and acquire and maintain

41 Makawao

Surveyed by Theophilus Metcalf, 1847. 56 x 56 cm.; Scale 1:9,504.
Source: Hawaii State Survey Division, Registered Map No. 185.

habits of enterprise and frugality, such as distinguish the common working population of England and the U. States?'

It is possible that this tract may have been *moku* land, or common land, as the concept was described earlier. By using *moku* land, no other claims by *ali'i* or *konohiki* would have to be resolved first.

This copy of a survey by Theophilus Metcalf bears no date, but undated annotations by an anonymous Government Survey staff member suggest it was made in 1847. According to other sources, the missionary Richard Armstrong made the surveys of the individual properties at the time of sale.[1] Metcalf's survey essentially includes the depiction of the land, or grants, that were sold in 1846. For each parcel, it shows the name of the owner, the Royal Patent number, and the acreage. Light annotations in pencil indicate Land Commission and Royal Patent numbers, which could not have been known until after 1850. It is possible that these annotations were made by C. J. Lyons in 1872 when he re-surveyed the area as one of the first projects of the Hawaiian Government Survey.

The pattern of properties shown here invite considerable speculation. There is reason to suspect that the properties do not conform to field or farm plots established under the original Hawaiian land tenure system. The individual parcels are much larger than *kuleana* lands awarded elsewhere in the islands, and the configuration of fields seems to have little if any correlation to local geography, especially stream patterns. It is noteworthy that one of the largest parcels, 98.93 acres awarded to J. S. Green, is labeled as a "model farm." Interestingly, there is no provision for roads or rights-of-way to gain access to any of the parcels.

Even though the sale of land to aliens was not approved until 1850, the map shows that several people with non-Hawaiian names owned land, including the five largest parcels, one of which was the "model farm."

42 Makawao, Maui

Surveyed by C. J. Lyons, 1872. 180 x 97 cm.; scale 1:12,000.
Source: Hawaii State Survey Division, Registered Map No. 603.

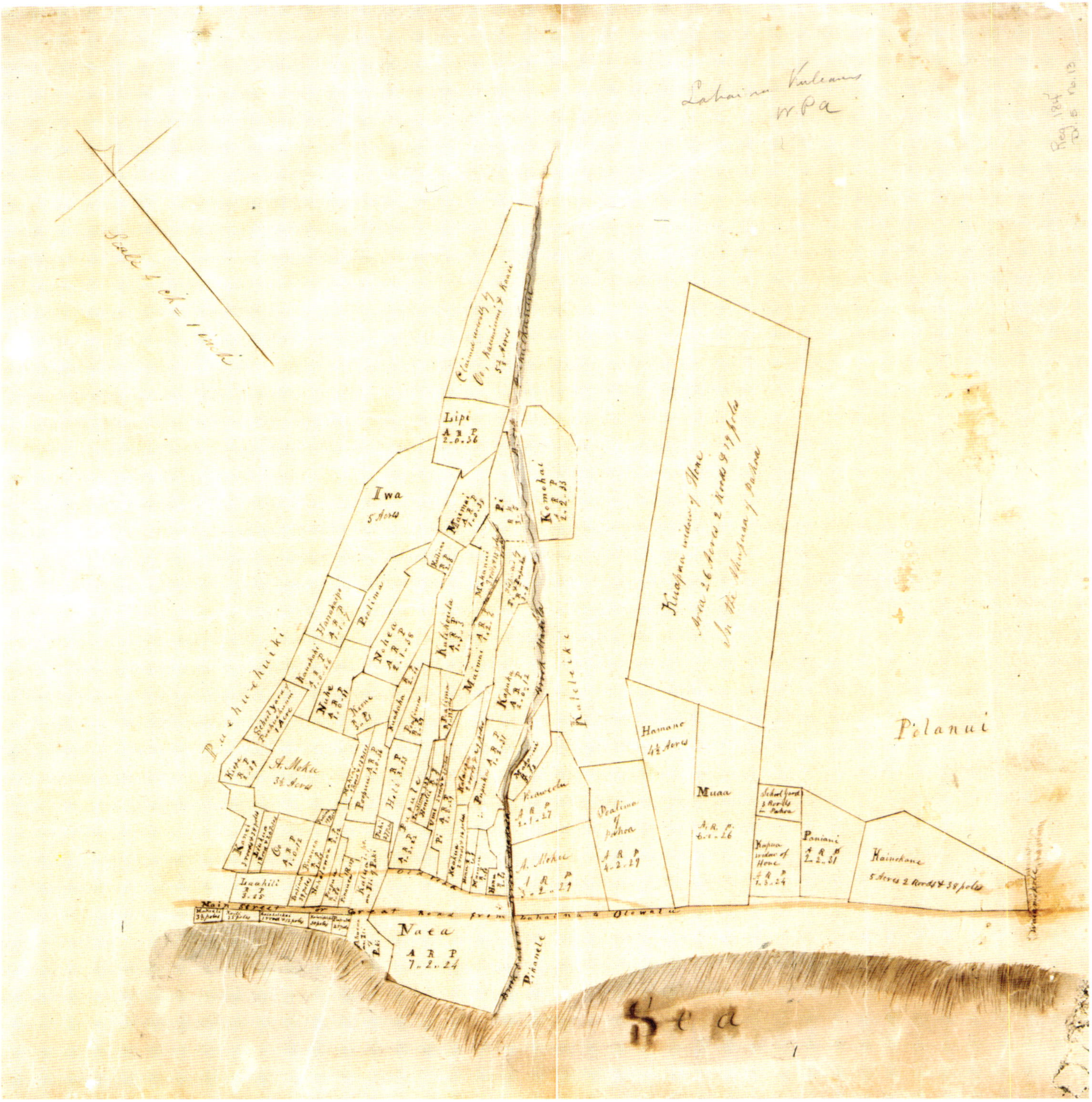

43 Lahaina

Surveyed by William P. Alexander, 1849 or 1850. 51 x 64 cm; scale 1:3,168 (4 chains to one inch).
Source: Hawaii State Survey Division, Registered Map No. 184.

44 Cartouche showing Lahainaluna

From: Kalama, Samuel P. *A map of the Hawaiian Islands according to the latest surveys*. Na Kalama i kaha. Lahainaluna: 1838.
Source: Geography and Map Division, Library of Congress; Hawaii—1838—Lahainaluna.

Lahaina

In the mid-nineteenth century, the bustling whaling port of Lahaina rivaled Honolulu in both size and importance. In 1794, Archibald Menzies, the surgeon and naturalist of Vancouver's expedition, described its congenial anchorage as follows:

> ... good with smooth water and well sheltered from the prevailing trade winds by the high mountains of Maui, and in great measure from all other accidental winds by the surrounding group of islands, viz: Molokini, Kahoolawe, Lanai, and Molokai, none of which are at a very great distance. It likewise possesses the advantage of an abundant supply of fresh water and every kind of vegetable refreshment, but hogs at this time were scarce.

At the time Alexander compiled his survey, Lahaina was a focal point for the whaling industry. A few years earlier, in 1846, a local census indicated that Lahaina consisted of 1,096 houses and 3,557 residents. The census was extremely detailed, noting that there were, among other things, 882 grass houses, 155 adobe structures, and 59 of stone or wood. The town also was home to 528 dogs.

This survey covers a delta-like area formed by the *'auwai*, or drainage ditches, coming out of the stream identified as Makila Brook on Alexander's survey. Areas watered by such well-developed irrigation systems supported a large number of people, and consequently many *kuleana* claims were submitted to the Land Commission.

The great agricultural productivity of this area may account for the presence of many small *ahupua'a*, some of which are less than eight acres in size. A delta-shaped region approximately one mile in length and a mile wide at the base contains portions of six *ahupua'a*: Pūehuehu Iki, Pūehuehu Nui, Kaulalo, Mākila, Pāhoa, and Polanui.

As an instructor at the missionary seminary at Lahainaluna, Alexander was intimately familiar with this area. Although this survey is undated, there is reason to believe it was done in 1849. That year, as noted in Chapter 4, he was chastised by the American Board of Commissioners for Foreign Missions for doing secular work—surveying—on Saturdays. This map appears to have been an appropriate weekend project for someone who taught surveying. Most likely, its purpose was to show the relationship of the *kuleana* lands in this densely cultivated area.

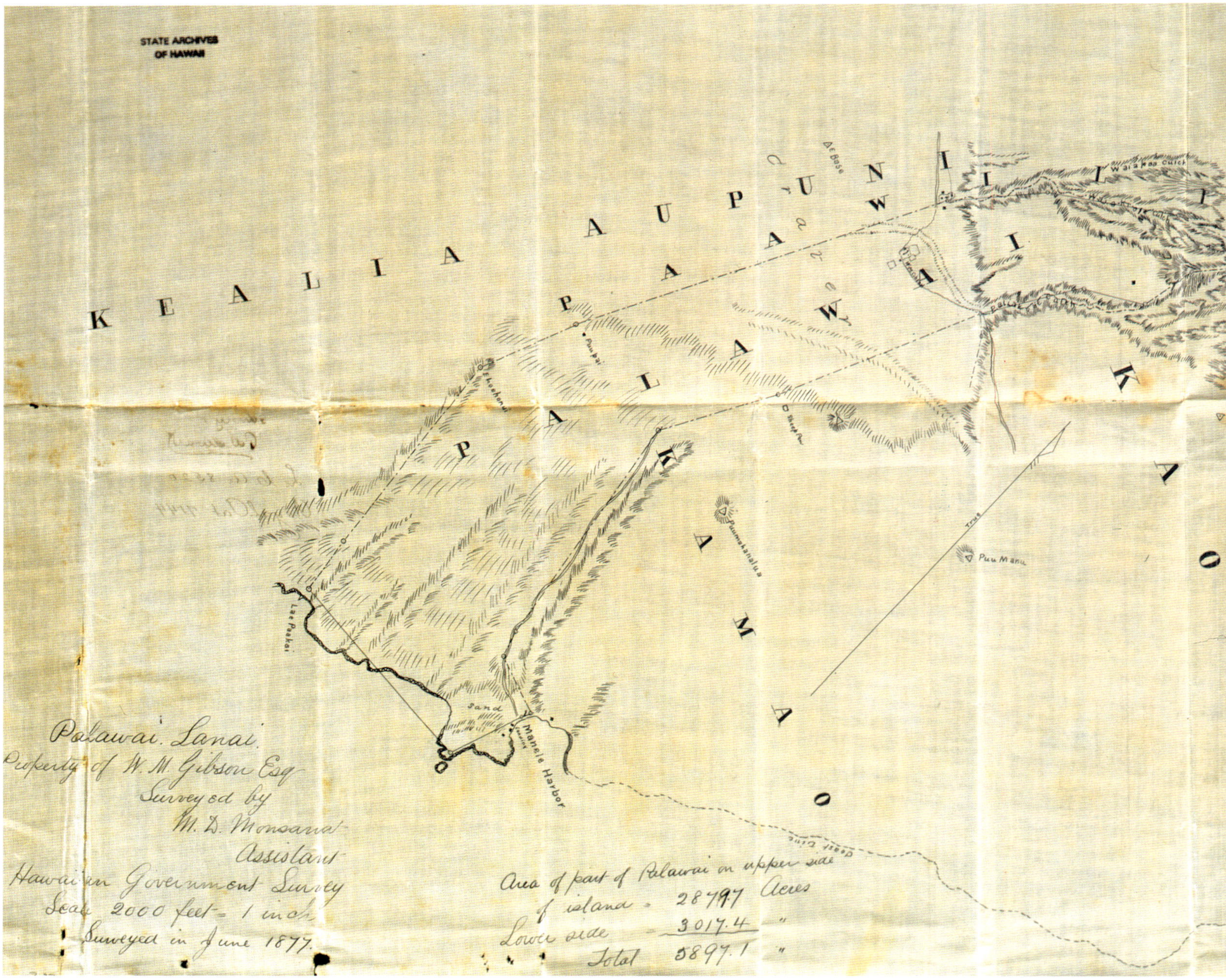

45 Pālāwai

Surveyed by M. D. Monsarrat, 1877. 39 x 77 cm.; Scale 1:24,000.
Source: Hawaii State Archives; G4382.L3:2P25 .H38.

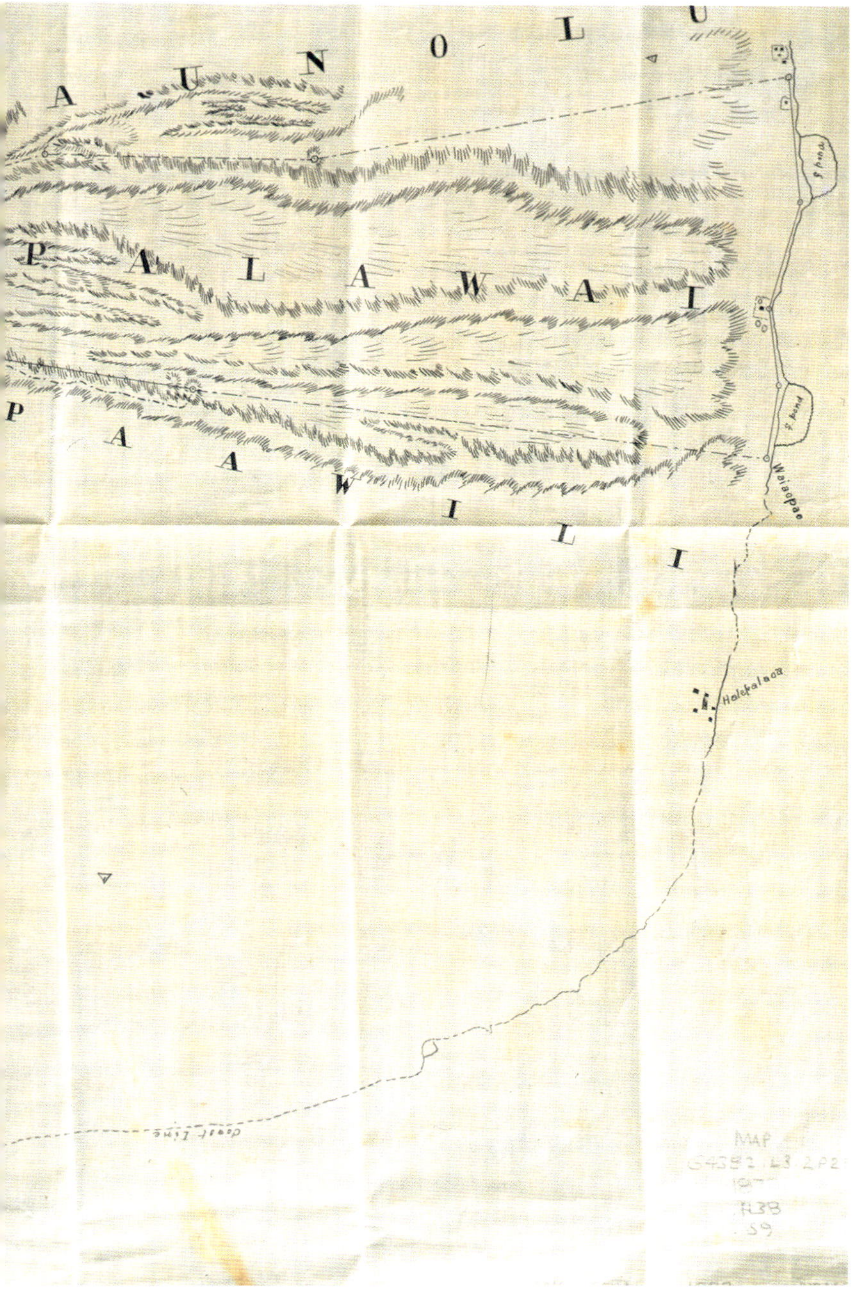

Pālāwai

The island of Lāna'i is unique in that some of its *ahupua'a* run clear across the island—from one coast up and over the mountain to the coast on the other side. Pālāwai, about ten miles long and one mile wide, is one such *ahupua'a*. In the *mahele* with Kauikeaouli in January of 1848, Pālāwai was awarded to Mikahela Kekau'ōnohi, the granddaughter of Kamehameha, as *apana* fourteen of LCA 11216, RP 7093. As was usually the case, the lands awarded in the *mahele* were awarded in name only—without a survey. It was not until 1877, when a subsequent owner wanted to receive a Royal Patent, that a survey was made.

In 1861, Walter Murray Gibson arrived in the islands and leased Pālāwai from Levi Ha'alelea for five hundred dollars as a gathering place for the Mormons in Hawai'i. By the end of 1863, Gibson had persuaded Ha'alelea to sell the *ahupua'a* to the Church. When the Church asked for the title, Gibson refused to hand it over and left the Church to develop Pālāwai as his own property.

Gibson asked the government for a Royal Patent, and there apparently was a question about the boundaries. The *ahupua'a* was surveyed in June 1877 by M. D. Monsarrat, who identified twenty-six corners and calculated that Pālāwai contained 5,897.1 acres. Monsarrat's survey, along with testimony by Gibson and Hawaii's Surveyor General, W. D. Alexander, was presented to William L. Moehonua, Commissioner of Boundaries in the 2nd Judicial Circuit on Maui. He found in favor of Gibson on September 29, 1877.

Monsarrat's survey of Pālāwai was incorporated into the general map of Lāna'i done for the Hawaiian Government Survey by Monsarrat and J. F. Brown in 1878.

The *ahupua'a* of Pālāwai begins on the south shore of Lāna'i and includes the west half of Mānele Bay and also Hulopo'e Bay, then runs north and northeast across the side of Pālāwai Basin, then up to the top of Lānai'hale at 3,300 feet and down the northeast side to the shore between Halepalaoa and Wai'ōpae. Gibson's ranch is shown in Pālāwai Basin, on the road from Mānele Bay to the little village of Kō'ele. By 1877, Gibson had left the ranch in charge of his children and had moved back to Honolulu to pursue a career in politics, ending up as Kalākaua's prime minister.

Lāna'i is generally quite dry, lying as it does in the rain shadow of Maui, and was never heavily populated. At the time of the *mahele*, there were perhaps six hundred people on the island, but when Monsarrat surveyed the area the population had dwindled to two hundred. Gibson was never able to develop Pālāwai the way he wanted, and it later passed into the hands of the Hawaiian Pineapple Company.

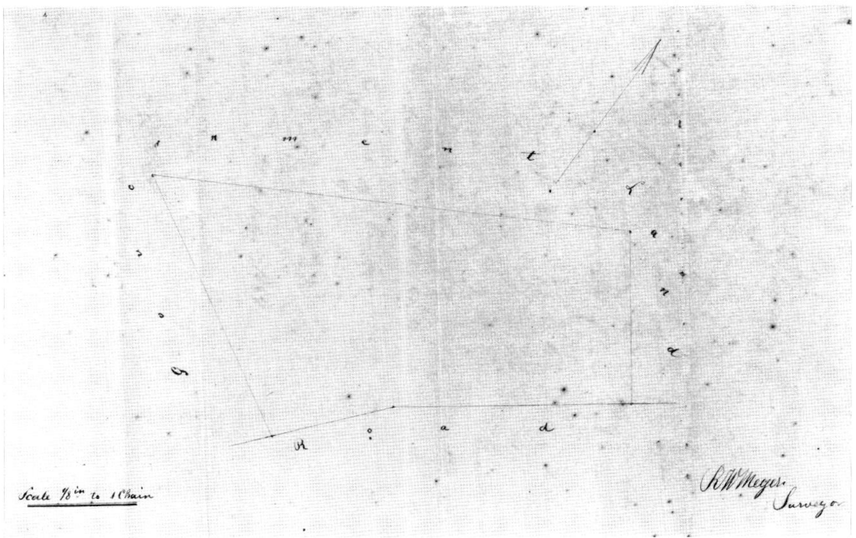

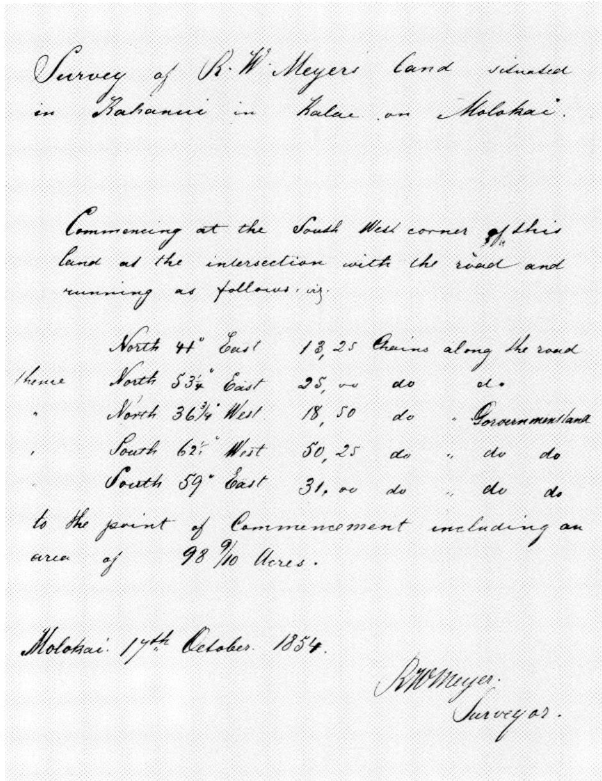

Survey of R. W. Meyers land situated in Kahanui in Kalae on Molokai.

Commencing at the South West corner of this land at the intersection with the road and running as follows, viz.

	North 44° East	18,25 Chains	along the road
thence	North 53¼° East	25,00 do	do
"	North 36¾° West	18,50 do	" Government land
"	South 62½° West	50,25 do	" do do
"	South 59° East	31,00 do	" do do

to the point of Commencement including an area of 98 9/10 Acres.

Molokai. 17th October. 1854.

RWMeyer
Surveyor.

46 Kala'e survey and description

Surveyed by Rudolf Wilhelm Meyer, October 17, 1854. 19 x 31 cm.; scale 1: 6,336 (8 chains to 1 inch).
Source: Hawaii State Bureau of Conveyances, Grant 724.

47 Kala'e description

Textual description accompanying Meyer's survey of October 17, 1854.
Source: Hawaii State Bureau of Conveyances, Grant 724.

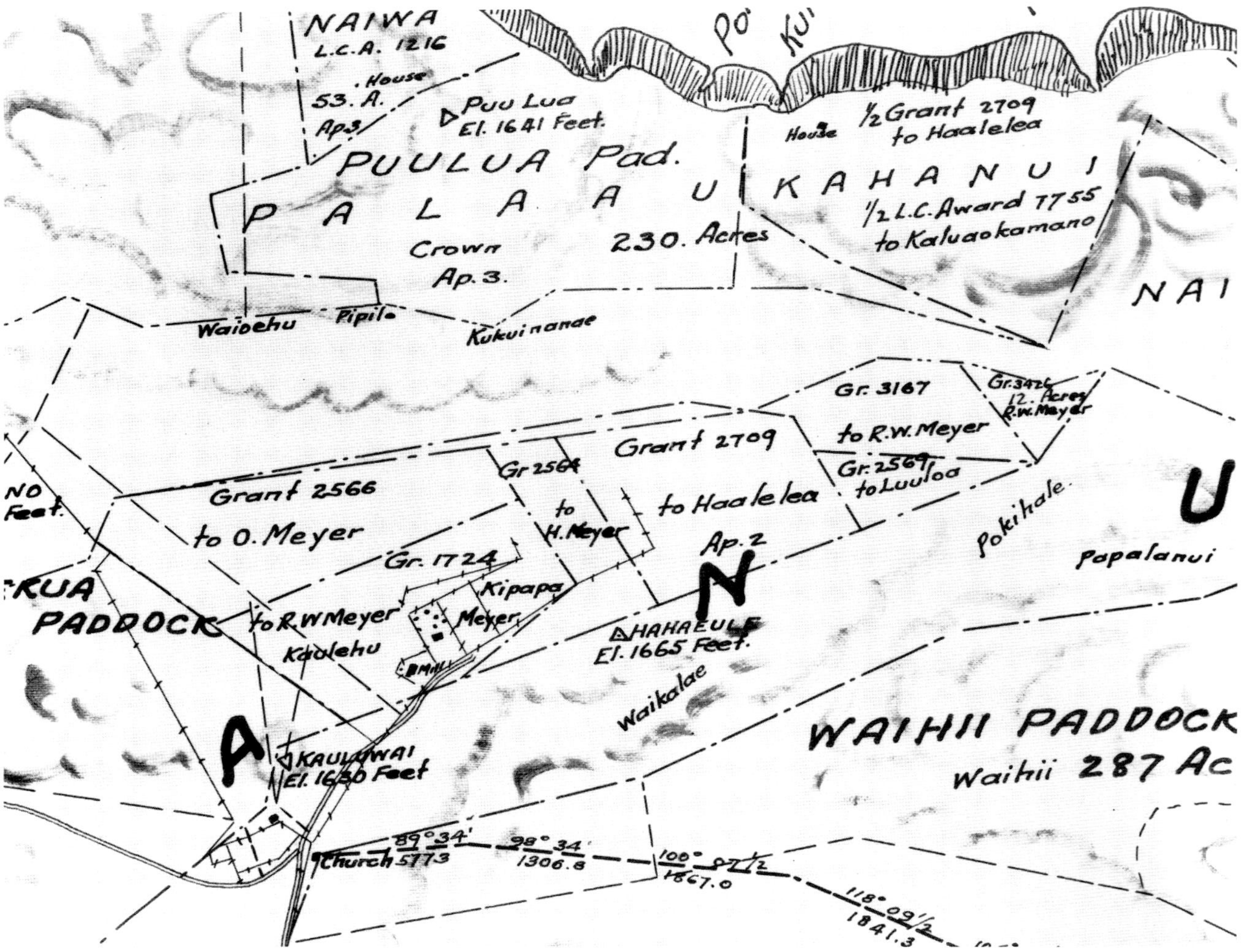

48 Moloka'i: Middle and West Section

1886. Surveyed by M. D. Monsarrat, 1886. 77 x 184 cm. scale 1:24,000.
Source: Hawaii State Survey Division; reg. map no. 1288.

49 Photograph of Kala'e Mill
Source: Molokai Museum and Cultural Center, R. W. Meyer Sugar Mill, Kala'e, Moloka'i.

Kala'e

Kala'e is situated in the *ahupua'a* of Kahanui on the island of Moloka'i. Kahanui is an unusual *ahupua'a* in that it consists of three unconnected *apana*. One is a small parcel on the south coast of the island at the mouth of the Manawainui Gulch, which includes the 'Ō'ō'ia fishpond. The largest *apana* is in the *kula* lands above the town of Kaunakakai and is remarkable for its extremely irregular shape. A third *apana* is approximately two miles east of the major *apana* and borders the heavily forested crest of the mountains of East Moloka'i.

As noted in Chapter 4, Meyer worked as a surveyor in the Honolulu area after arriving in the islands in 1850. In October 1850, he moved to Moloka'i to work as a government surveyor, and in 1851 married a Hawaiian *ali'i* named Dorcas Kalama Waha. Although Kala'e' is located in cool, moist *kula* lands at an elevation of about fifteen hundred feet—a very bountiful area—John Wesley Coulter's maps indicate that no one lived in the area at the time Meyer purchased it. This initial purchase from the government encompassed 98.9 acres, and is listed as Grant 1724. The title indicates that Meyer paid one dollar fifty cents per acre.

He enlarged the estate by purchasing additional parcels. In 1859, he bought 31.18 acres under the name of his young son, Henry, identified as Grant 2564, and 85.86 acres for another young son, Otto, identified as Grant 2566. In his own name, he purchased 57.1 acres in 1877 (Grant 3167), 12 acres in 1888 (Grant 3426), and 1,048 acres in 1888 as grant 3437. All parcels were situated in the same *ahupua'a*.

Meyer was noted for the diversity of his enterprise. In addition to cattle ranching, he raised coffee, corn, potatoes, sugar cane, wheat, fruit, mulberry trees, and grapes, and kept silkworms, beehives, and dairy cattle. A sugar mill, which is now a museum, was operated between 1878 and 1889.

This survey of Kala'e is extremely rudimentary. There is no reference to the location of the initial point or to whether the north arrow is pointing to true or magnetic north. The textual description of the parcel is given on a separate page.

Alanui Beritania
King St
Queen St
Alanui Alii
Alanui Kalepa
Alanui Hotele
Ala Kea
Alanui Kahuna
Alanui Richards
Alanui Nuuanu
Alanui Papu
Kaeo
Kaliu
Kahawai o Makaho

51 No. 2 View of Honolulu, From the Catholic Church

Drawn by Paul Emmert.
Source: Hawaii State Archives; no. 16.653.

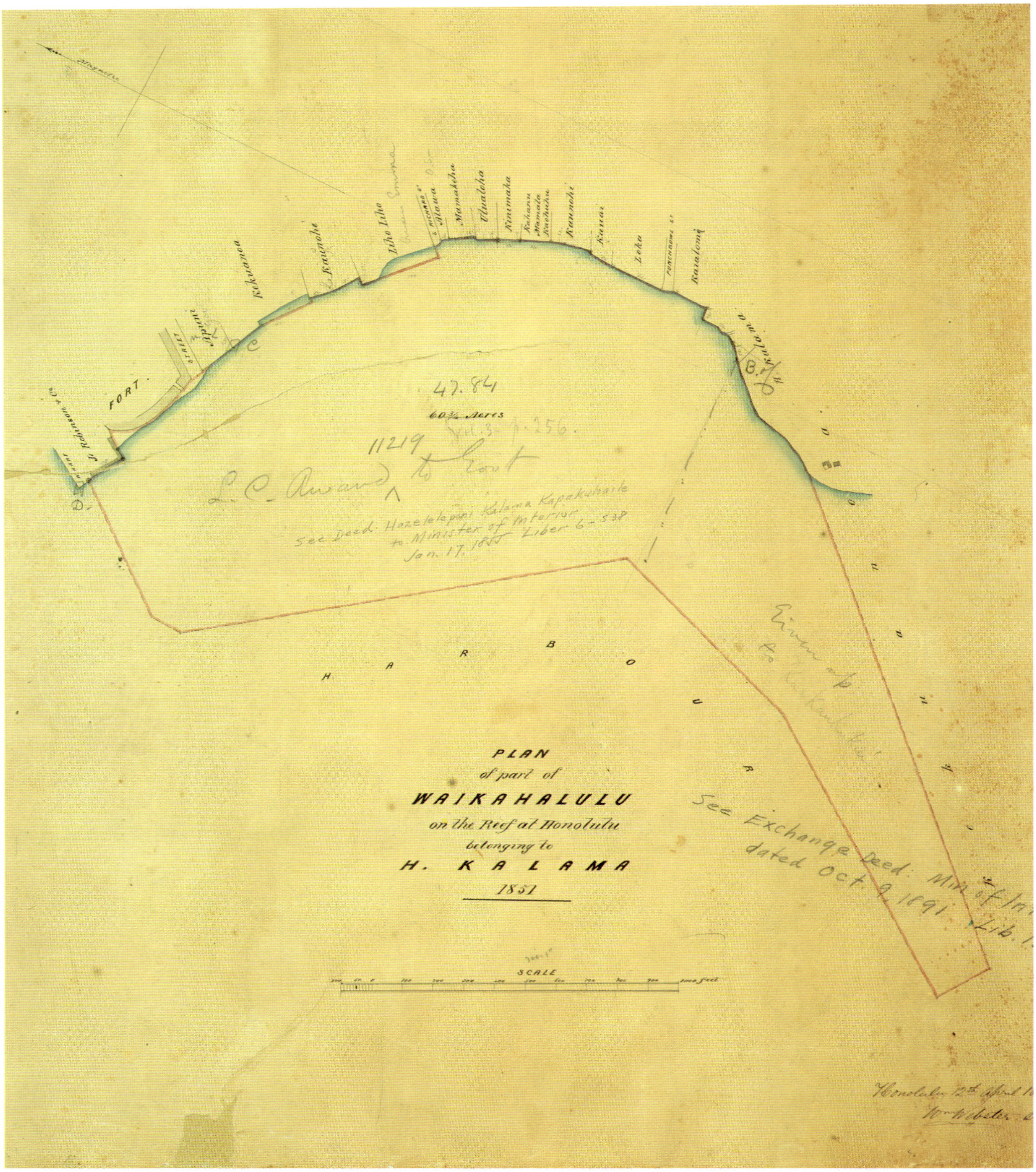

52 Waikahalulu, O'ahu

Surveyed by William Webster, April 12, 1851. 51 x 50 cm.; scale 1:2,400 (200 feet to one inch).
Source: Hawaii State Survey Division; reg. map no. 244.

Waikahalulu, O'ahu

Honolulu emerged as the focal point of political and economic activity in Hawai'i because it was the favored port of European vessels visiting the islands. Honolulu harbor was well-protected and deep. It was not, however, particularly big in its natural state. As early as 1844, a plan had been put forth by Robert Crichton Wyllie, then the British consul but later the Minister of Foreign Affairs for the Kingdom of Hawai'i, to improve the harbor itself as well as the onshore facilities. Nothing was done about the suggestions at that time.

By 1850, however, the increased shipping business in Honolulu required better facilities, and new developments in Pacific trade patterns offered the hope that Honolulu might be a focal point for much of the trans-Pacific steamer trade. The government responded with a plan to improve the harbor. An ambitious design that would have improved both sides of the harbor was abandoned because of financing problems and other factors. Efforts to improve the eastern side did commence, however.

William Webster was hired to design and survey the proposed improvements, the goal of which was to extend the wharf by building a wall along the Waikahalulu reef to the southeast of the harbor and filling the area between the wall and the shore.

The area to be filled was offshore from the *'ili* of Waikakahulu, which had been given to Queen Kalama during the *mahele*. Kalama claimed the Waikahalulu Reef as an integral part of her land, arguing that any filled area would also belong to her. Without legal precedence on which to resolve the dispute, it went to the Hawaiian court. Finding that her right to the reef extended only to fishing and stones that might be cut from it, the court decided in favor of the government. Kalama resisted, however, and the government finally settled with her in 1854 by paying $22,000 for title to the land, which was finally received on January 17, 1855. Webster's survey shows the shoreline as it existed in 1851 in blue, the area that was to be filled in red, and the names of owners of the waterfront properties set off perpendicular to the shoreline.

The filled area was subdivided for sale to the public and became known as the "water lots." A survey of these lots was executed in October 1857 in anticipation of the sale. Although there is no credit for the surveyor, the distinctive drafting, coloring, and bar scale suggest that it was also the work of William Webster. Among the noteworthy details on the survey are the delineation of the new proposed streets, although the east-west streets are not named. The new lots, which are outlined in pink, are designated by two different names: One group of ten parcels is identified as the "rear lots," and the other seven are specifically called the "water lots." Aloha Tower, the landmark of modern Honolulu Harbor, sits on lots 1, 2, and 3 of the "water lots" highlighted in pink. It should be noted, however, that the street pattern that ultimately emerged for this area did not follow the plan shown on this survey, partly because it took so long to complete the project. At least into the 1880s, the development had extended only about half as far to the east as proposed in the 1851 survey, and it was not until 1893 that maps show the project had been completed.

The inset chart of Honolulu Harbor published on the 1857 nautical chart of southern O'ahu issued by the British Admiralty was originally surveyed by Lieutenant Charles R. Malden in 1825 (EMH Ill. 33). Master Hull updated Malden's plan of Honolulu to reflect the latest developments in 1856. The fort had been torn down and its stones used for the retaining wall shown running along the edge of the reef in anticipation of being filled with dredged material from the harbor. The new prison to house the prisoners formerly held in the fort is depicted in Iwilei. Added to this copy in red ink is a location of a proposed lighthouse to help guide ships up the narrow channel into Honolulu Harbor. The lighthouse that was eventually built rises from the location designated "Red lanthorn."

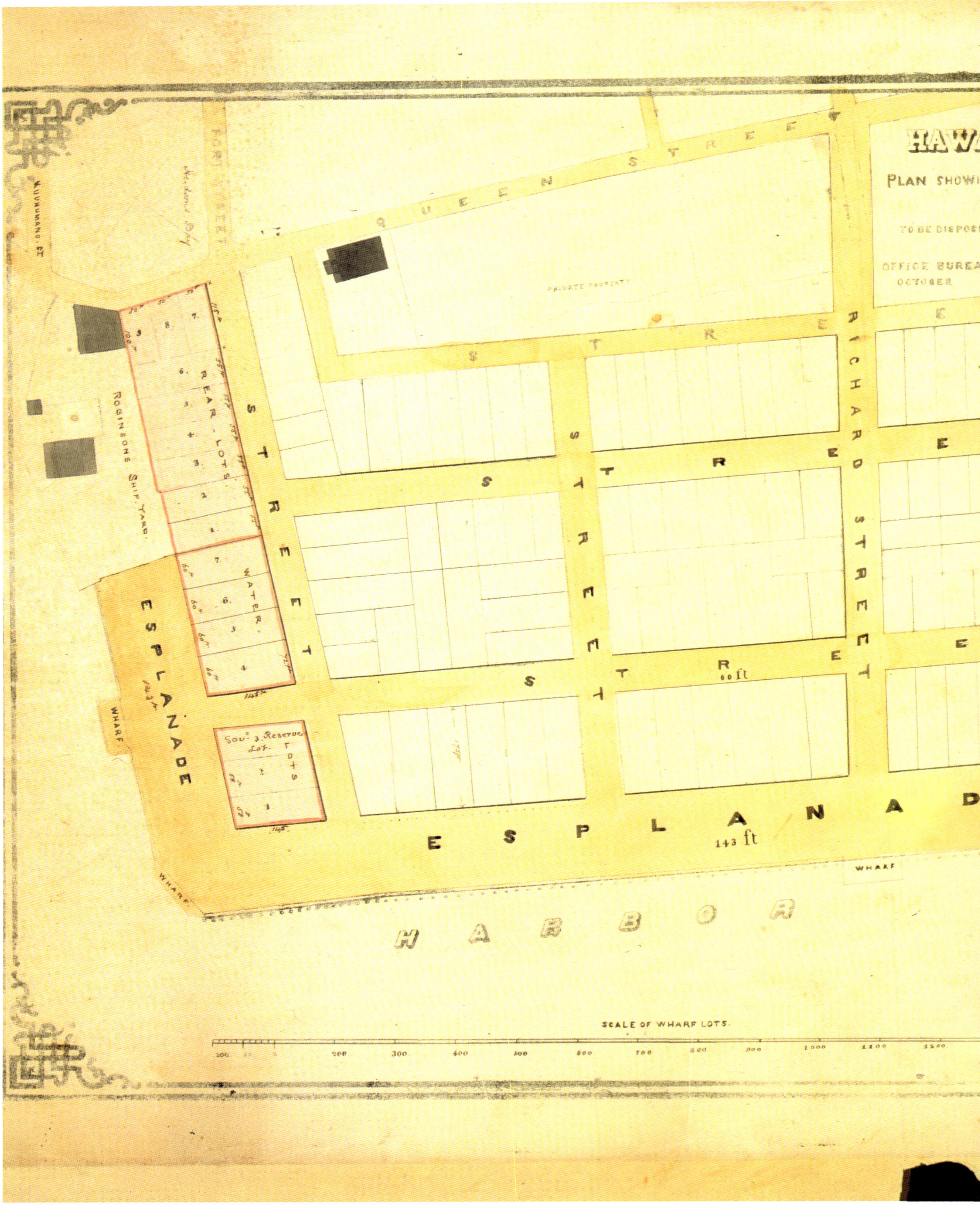
QUEEN STREET
FORT STREET
ROBINSONS SHIP YARD
REAR LOTS
WATER
PRIVATE PROPERTY
RICHARD STREET
STREET
ESPLANADE
WHARF
60 ft
143 ft
HARBOR
SCALE OF WHARF LOTS

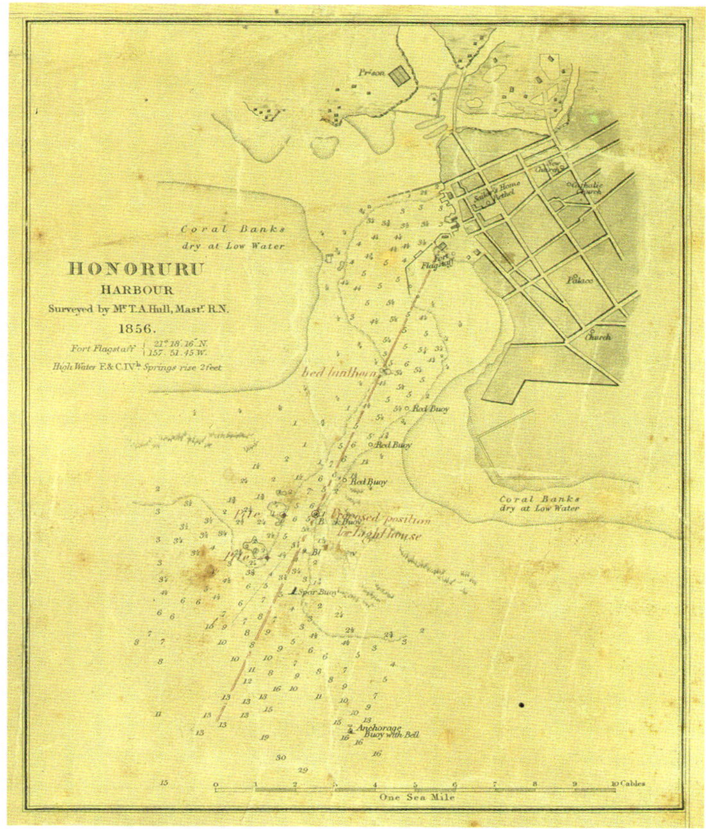

53 Honolulu Harbor

Surveyed by T. A. Hull, 1856. 20 x 16 cm.; scale ca. 1:20,000.
Source: Library of Congress, Geography and Map Division; Charts—H.O. 1378

54 Waikahalulu Water Lots, O'ahu

Surveyed by William Webster, October 1857. 50 x 64 cm.; scale 1:1,200.
Source: Hawaii State Survey Division; reg. map no. 279.

55 Mānoa Valley, ca. 1870
Source: Hawaii State Archives; neg. no. 1991.085.

Mānoa Valley and the ʻili *of Wailele*

Because it is somewhat more complex than the usual land transaction in the *mahele*, the *ʻili* of Wailele in Mānoa Valley, Oʻahu, makes an interesting case study. For one thing, the story of Wailele shows that gifts of land and grants made as early as the time of Kamehameha were honored during the *mahele*.

Wailele lies near the mouth of Mānoa Valley, just *mauka* and *ewa* of the University of Hawaiʻi. Some of the land is still open, preserved as athletic fields for the Mid Pacific Institute, a private school that now occupies the site.

The story of Wailele begins with Captain George C. Beckley, an English navigator who came to the islands sometime before 1806 and became one of Kamehameha's military advisors. He married an *aliʻi* named Ahia and was himself made a high chief by Kamehameha in 1815. When Kalanimōkū, Kamehameha's chief advisor, and John Young completed the Honolulu fort in 1816, Beckley was given command of the fort and also received the *ʻili* of Wailele. He died in 1825.

On March 7, 1827, Boki, the governor of Oʻahu, leased the Wailele property to British consul Richard Charlton for a period of twenty-five years, but with the understanding that afterwards it was to return to the Beckley family. Charlton, one of the most nefarious foreigners in all of Hawaiian history, later started a bitter and long-running diplomatic dispute between Hawaiʻi and Great Britain when he contested the extent of lands in Honolulu that he claimed had been leased to him for 299 years. In a claim against the Hawaiian government, the land at Wailele was mentioned.

To avoid devoting several pages to an explanation of the complicated and nasty Charlton affair, it can be stated simply that Charlton's lease was transferred to Louis Desire Maigret, the pioneer Catholic priest and future Bishop of Hawaiʻi, in 1845. This lease was the first recorded instrument in the books of the Registry Office. It is in French and is dated November 6, 1844. The property reverted to the Beckley heirs in 1855.

After the registration of claims began during the *mahele*, the executor of the Beckley estate and legal guardian of the children, Alexander Adams, took steps to preserve the Beckley claim to Wailele. First, on April 20, 1847, Theophilus Metcalf was hired to survey Wailele (Ill. 58). As part of the process, the Beckley family was required to provide testimony as to the validity of their claim, for which the following was given to the Land Commission on October 22, 1847.

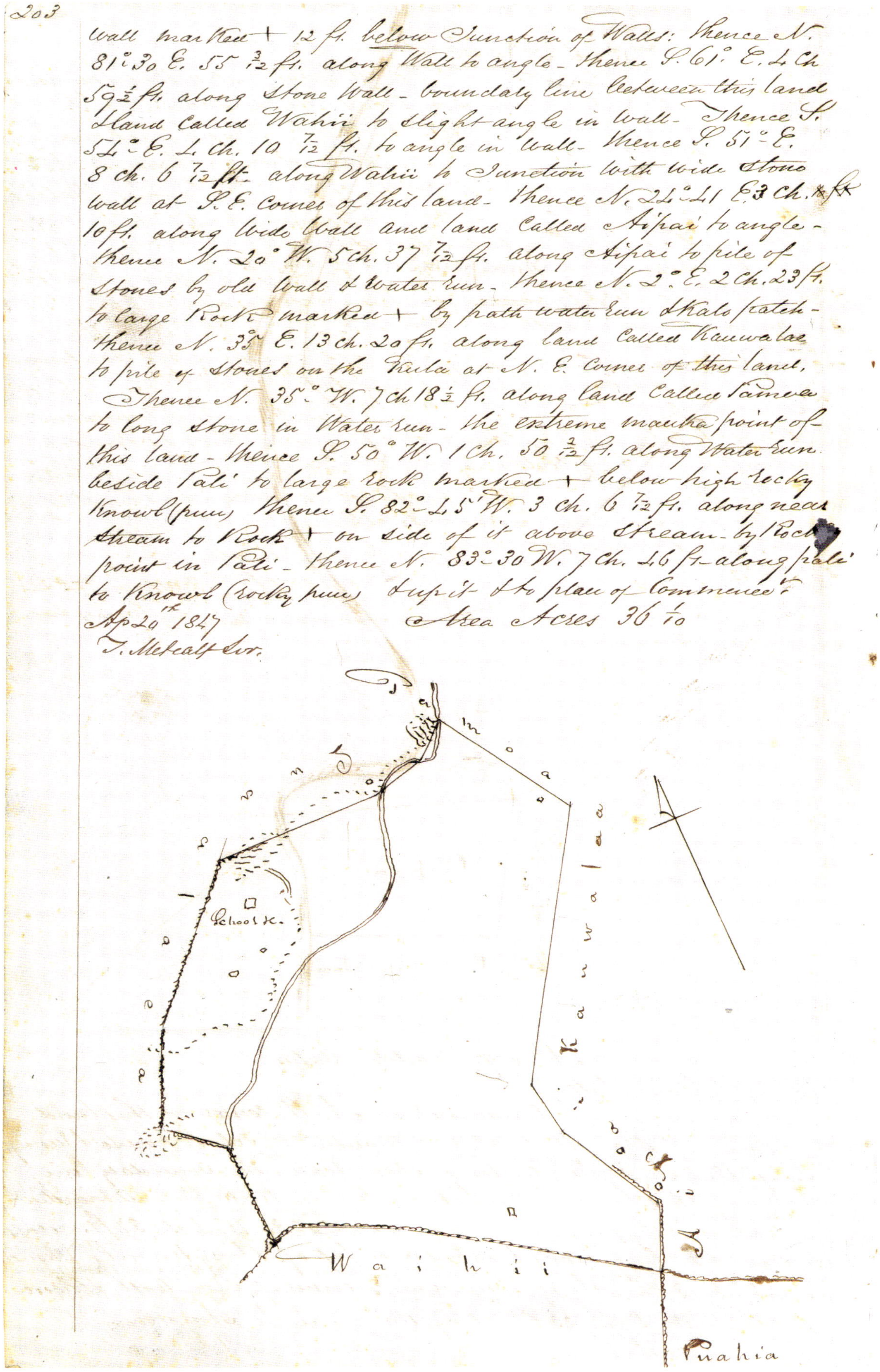
203
wall marked + 12 ft. below Junction of Walls: thence N.
81°30 E. 55 ½ ft. along Wall to angle - thence S. 61° E. 4 Ch
59½ ft. along Stone Wall - boundary line between this land
& land called Wahii to slight angle in wall - Thence S.
54° E. 4 Ch. 10 ½ ft. to angle in wall - thence S. 51° E.
8 ch. 6 ½ ft. along Wahii to Junction with wide stone
wall at S.E. Corner of this land - thence N. 24°41 E. 3 Ch. ×ft
10 ft. along wide wall and land called Aipai to angle -
thence N. 20° W. 5 Ch. 37 ½ ft. along Aipai to pile of
stones by old wall & water run - thence N. 2° E. 2 Ch. 23 ft.
to large Rock marked + by path water run & kalo patch -
thence N. 35° E. 13 Ch. 20 ft. along land called Kauwalae
to pile of stones on the Kula at N. E. Corner of this land.
Thence N. 35° W. 7 ch 18½ ft. along land called Samea
to long stone in water run - the extreme mauka point of
this land - thence S. 50° W. 1 Ch. 50 ½ ft. along water run
beside Pali to large rock marked + below high rocky
knowl (puu) thence S. 82° 45' W. 3 ch. 6 ½ ft. along near
Stream to Rock + on side of it above stream by Rock
point in Pali - thence N. 83°30 W. 7 Ch. 46 ft. along pali
to Knowl (rocky puu) & up it & to place of Commencement.
Ap 20th 1847 Area Acres 36 1/10
T. Metcalf Svr.

56 Plat and textual description of Wailele, 1847

Surveyed by Theophilus Metcalf, April 20, 1847. 31 x 20 cm.; scale not given.
Source: Hawaii State Archives; *Land Commission Awards*, No. 819.

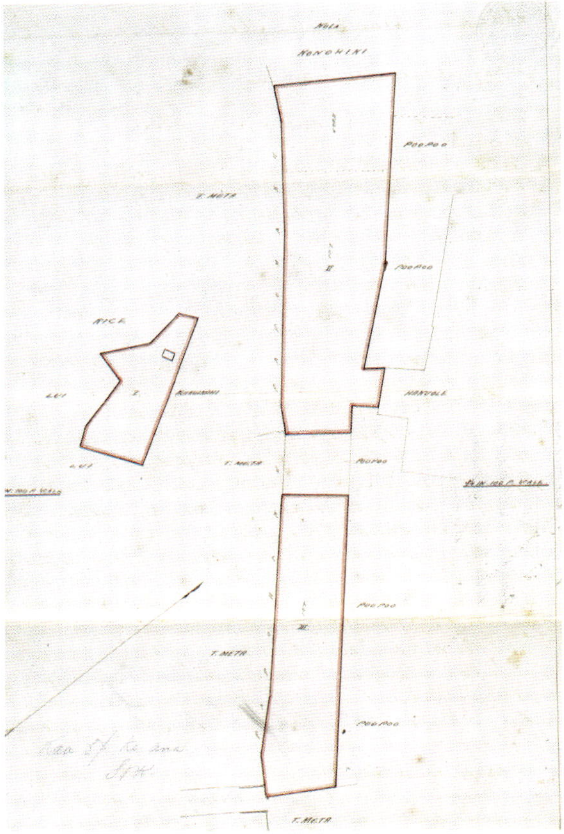

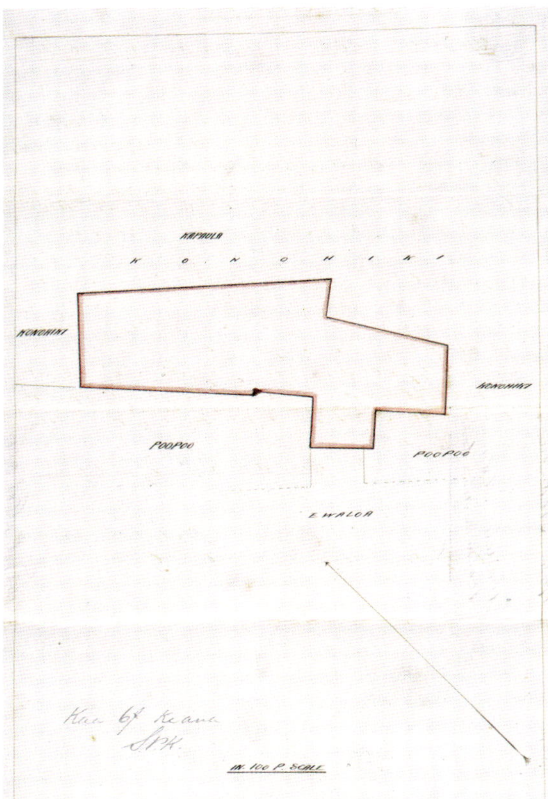

57 Ewaloa Kuleana

Surveyed by S. P. Kalama, Sept. 26, 1854. 30 x 20 cm.; scale 1:900.
Source: Hawaii State Archives; *Land Commission Awards*, Royal Patent 3031.

58 Hakuole Kuleana

Surveyed by S. P. Kalama, Sept. 26, 1854. 30 x 20 cm.; scale 1:1,200.
Source: Hawaii State Archives; *Land Commission Awards*, Royal Patent 2271.

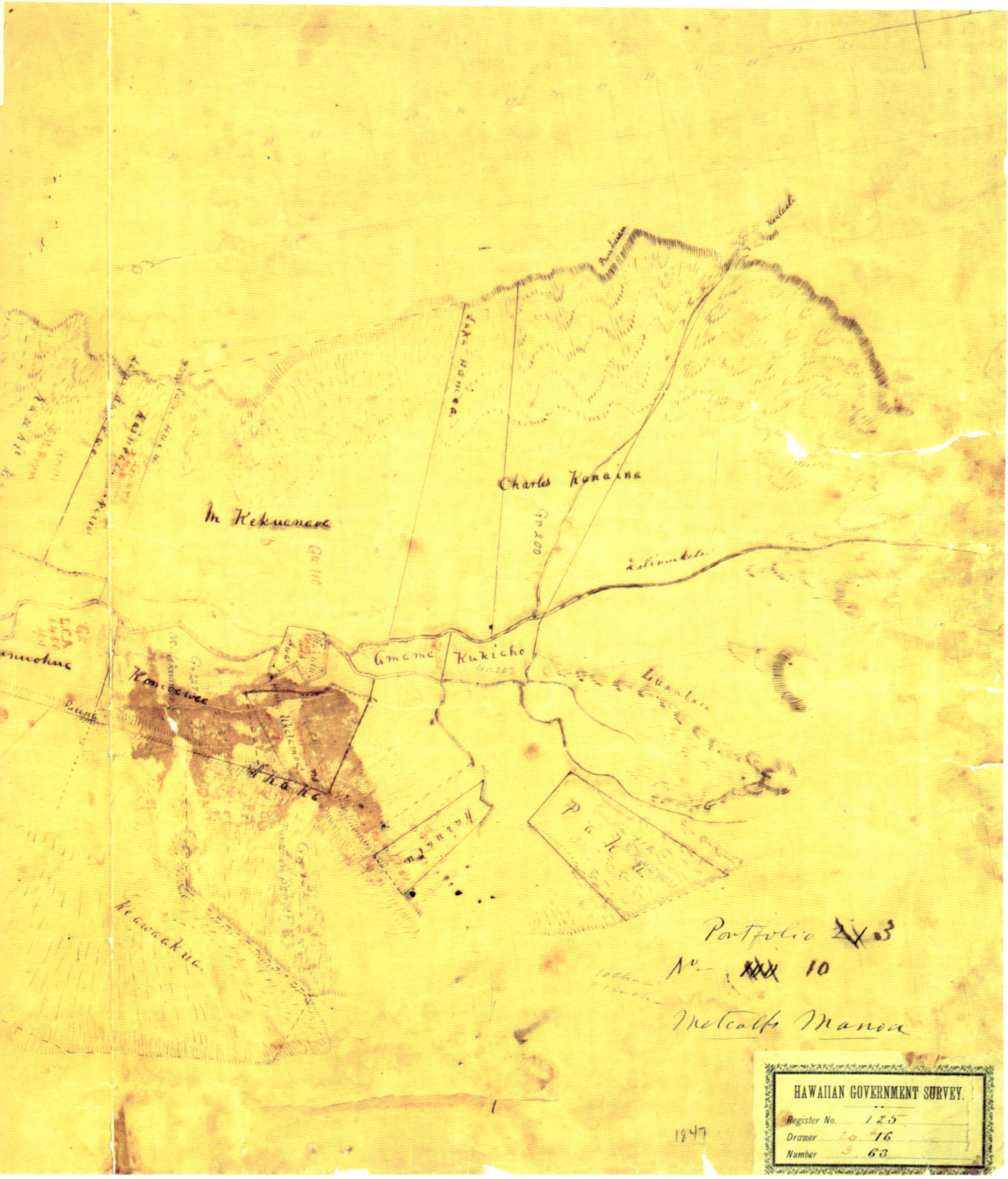

59 Mānoa Valley, 1847

Surveyed by Theophilus Metcalf, 1847.
51 x 71 cm.; scale 1:7,920.
Source: Source: Hawaii State Survey Division; reg. map no. 125.

Also to our farm called Wailele situated at the mouth of the Valley of Manoa, Is. of Oahu, sold to the said George Beckley by Kalanimoku (then Gov. of Oahu) about or in the Year 1815 and was in the uninterrupted possession of the said George Beckley and his heirs until 1827, when on the 7th day of March 1827, said farm was let to Richard Charlton by Poki, governor of Oahu, for the term of 28 years, to be returned to the heirs of the said George Beckley at the expiration of said term.[2]

Additional testimony was collected on June 19, 1848:

The heirs of George Beckley viz: Ahia (widow of Beckley), William Beckley, Maria Beckley (wife of Kamakahonu), Abram (infant child of Nancy Beckley), Mary Beckley (wife of A. H. Fayerweather), George Frederick Beckley (absent from the Islands since 1834 and last heard from in 1841 at Valparaiso), Marie Beckley (wife of Geo. W. Punchard).

A. H. Fayerweather, as attorney for the heirs, stated that the lands comprised in these three claims, were given by the King (at the late division of lands) to the heirs of Beckley without any division, and that he had presented a certificate to this effect to the President of the Board (which could not be found: case postponed until another can be obtained). He further stated that in the late division, the King left these lands untouched and took three other lands on Hawai'i called "Muliwae" in Waipio, "Waawaa" or Kipapai, and "Pualoalo" on Kohala.[3]

As noted in Chapter 3, claims had to be filed between February 14, 1846, and February 14, 1848. The Land Commission did not get around to acting on those claims until later. The Beckleys finally received their Land Commission award to the *'ili* of Wailele on March 29, 1852. An annotation on the award, however, noted that it was made "preserving internal rights of natives." In short, the award was not for all of the *'ili* of Wailele, but only that part which was not claimed by *maka'āinana* as *kuleana* lands.

Metcalf's survey shows the names of the property owners surrounding Wailele, which is useful and necessary information for relating it to other lands in the area, but he does not show that there were five pieces of property within Wailele for which *kuleana* claims had been filed. It was implicit in all claims such as the Beckleys' that *kuleana* claims had to be honored, hence the annotation on their award.

Five individuals had indeed filed claims for *kuleana* lands in Wailele within the required time period. A person named Ewaloa received 2.39 acres, Hakuole received .75 acres, Hoohoku received .95 acres, Paikau received 1.3 acres, and Wahahee received .89 acres. As required by the Land Commission, these people also had surveys made of their lands. The Hakuole survey, bearing the initials of S. P. Kalama, shows that it was a very conventional property, consisting of only one *apana*, or part (Ill. 60). As shown in another survey initialed by S. P. Kalama, the Ewaloa award consisted of three unconnected *apana* (Ill. 59). The three *apana* are not shown in their correct geographic relationship to each other, however.

In 1860, for reasons unknown, a survey was made of the Wailele property by W. H. Pease (Ill. 62). Although Pease does not show the boundaries of the five *kuleana* lands within Wailele, he does note on the survey the names and acreage of the claims.

On March 7, 1855, the lease to Maigret expired, and the Beckley heirs went again to the government. By paying their commutation of fifty dollars, based on the government's appraisal by John Fuller and John W. Makalena of one-fourth the value of the undeveloped land, the Beckley heirs received the final legal document in the claims process—Royal Patent 4546—on September 19, 1862, signed by the Minister of Interior, Lot Kamehameha.

As explained in Chapter 3, the government experimented with the sale of fee simple lots in Makawao on Maui and Mānoa Valley on Oahu, with John 'I'i acting as land agent. This experimental sale was reported in *The Polynesian* on September 2, 1848, and included the names of all the buyers of the lots. Theophilus Metcalf, the original surveyor of the Beckley property, produced a survey of Mānoa showing the parcel numbers and grant numbers of the thirty lots that were sold as well as the other lands already claimed within the valley, including Wailele. Metcalf's map represents some of the most difficult work that surveyors face, correlating independent plats of neighboring properties to identify vacant areas or lands for which there are conflicting claims. This is the first survey of that type that was done in Hawai'i.

Metcalf did not, however, show all the *kuleana* lands that were within the larger lands, such as Wailele. It was not until 1882, in a map compiled by Erdman Dwight Baldwin of the Hawaiian Government Survey, that both Wailele and the *kuleana* lands within it were shown together.

Even as Mānoa Valley was developed over the following decades, the *'ili* of Wailele remained a distinct entity in the landscape because it basically was kept together as a large parcel of land. In the photographs that date from the early twentieth century (Ill. 63) and the 1920s (Ill. 65), Wailele can be identified by two distinctive multi-winged, multi-dormer buildings—the Kawaiahao Seminary for girls and the Mills School for boys.

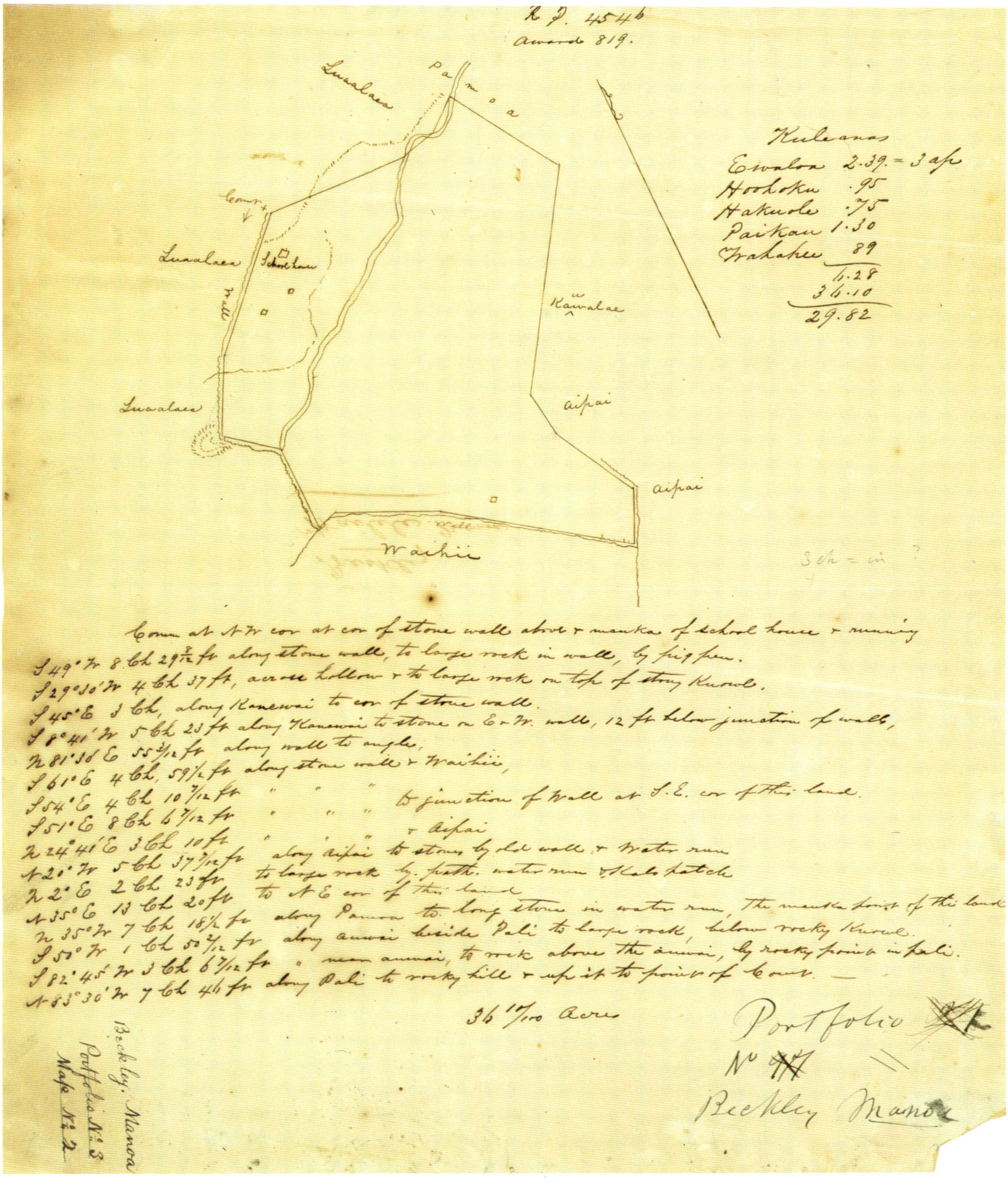

60 Survey of Wailele, 1860

Surveyed by William Howard Pease, 1860.
43 x 36 cm.; scale 1:2,376 (3 chains to 1 inch).
Source: Source: Hawaii State Survey Division; reg. map no. 78

61 Aerial photograph of Mānoa Valley, 1927. Wailele is in the lower left corner.
Source: Hawaii State Archives, negative no. 17,092.

62 Mānoa Valley, ca. 1920. Wailele is in right center.
Source: Ev Wingert.

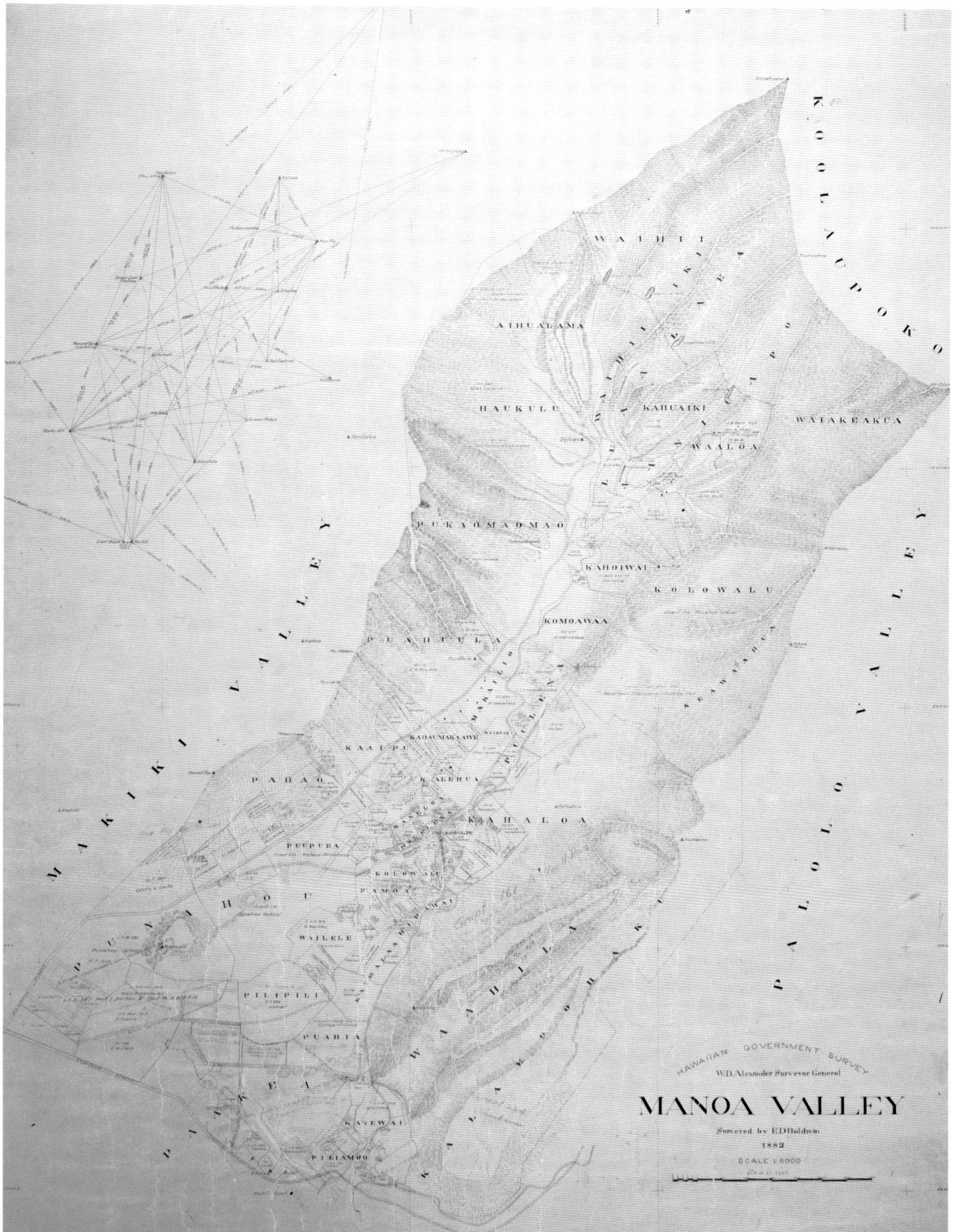

63 Map of Manoa Valley, 1882

Surveyed by Erdman Dwight Baldwin, 1882.
140 x 98 cm.; Scale 1:6,000.
Source: Hawaii State Survey Division; reg. map no. 1068.

64 Waikāne, O'ahu

Samuel P. Kalama, 1850.
46 x 50 cm.; Scale 1:15,840 (20 chains to an inch).
Source: Hawaii State Survey Division; reg. map no. 64.

65 Waikāne, from Mokuliʻi Island, 1993
Source: Gary L. Fitzpatrick

66 Waikāne Valley, Oʻahu
Source: Bernice Pauahi Bishop Museum, Honolulu, Hawaiʻi; negative number CP 54598.

Waikāne, Oʻahu

The windward shore of Oʻahu in the area of Kāneʻohe Bay is notable for a broad coastal plain, nearly three miles wide at some points, that is backed by the sheer *pali*, or cliff, of the Koʻolau Mountains. The *ahupuaʻa* of Waikāne is a valley type common to mountainous areas that receive heavy rain. The mouth of the valley at the shore is relatively narrow, about one mile across. Inland, however, it widens into a broad amphitheater that is nearly two miles across. It extends nearly three miles to the summit ridge of the Koʻolau, which reaches an elevation of approximately 2,700 feet.

Prior to the *mahele*, Waikāne belonged to Laura Konia, a granddaughter of Kamehameha. Her husband, Abner Pākī, was an *aliʻi nui* who controlled a large number of lands in Hawaiʻi. After the *mahele*, Waikāne was designated as government land. Records indicate that there were nineteen awards of *kuleana* within Waikāne, totaling 53.5 acres and ranging in size from .15 to 7.75 acres. Six *ʻili* within Waikāne had been awarded to the government in the *mahele* before the rest of the land ended up being designated as government land.

One grant for 1,698.5 acres in the *ahupuaʻa*, under Royal Patent 464, was sold to Edwin Oscar Hall and Henry Dimond in 1850, both of whom had come to Hawaiʻi as missionaries. In 1849, Hall left the mission and took the position of minister of Finance. He later founded a mercantile firm, and in 1873 became minister of the Interior. Dimond, a printer by training, left the mission in 1850.

Kalama's survey, which was made in 1850, shows none of the *kuleana* or *ʻili* within Waikāne. However, it is interesting in that it identifies a large area as taro lands. He also indicated the location of a church land as "*aina hale pule*" and another feature "*kahua*," a foundation or house site. According to census records, Waikāne contained only 117 inhabitants in 1849, a figure that seems low given the richness of the land. An explanation for this may lie in the comments of Robert Crichton Wyllie, the minister of Foreign Affairs at the time, who noted that roaming cattle belonging to the *aliʻi* and foreigners had destroyed more than one hundred acres of good crops and prevented the *makaʻāinana* from planting their crops. Having been forced from their fields by the cattle, many Hawaiians may have abandoned land that they would have been able to claim as *kuleana* grants.[4]

The rugged nature of the Koʻolau Mountains obviously prevented Kalama from making a personal survey of the entire boundary of Waikāne. Several prominent peaks along the borders of Waikāne are identified as corners of the boundary line, but the straight lines connecting them indicate that he did not actually survey the intervening segments. In fact, the configuration of the highest portions of the boundary of Waikāne on Kalama's survey does not fit well with the boundaries portrayed on modern topographic maps.

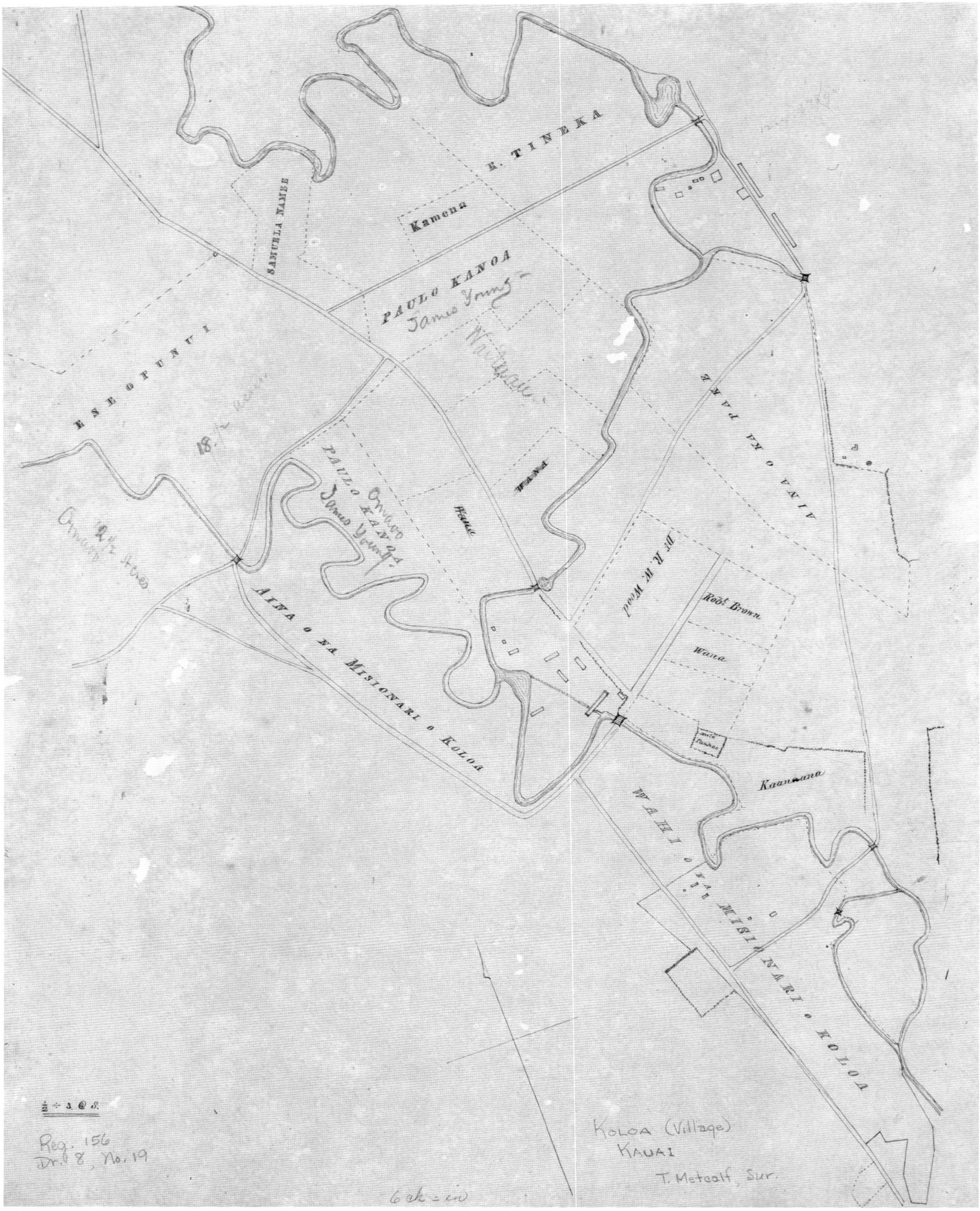

67 Kōloa, Kauaʻi

Surveyed by Theophilus Metcalf, 1847?
41 x 34 cm.; scale 1:4,752 (6 chains to an inch).
Source: Hawaii State Survey Division; reg. map no. 156.

Kōloa Plantation

The *ahupua'a* of Kōloa is on the southern coast of Kaua'i in an area of many modern hotels and condominiums now frequently referred to as Poipu. Kōloa plantation was the site of the first large-scale effort at sugar production in Hawai'i. In 1835, Kauikeaouli leased one thousand acres of land there to a firm known as Ladd and Company. A survey of the leased land either was not made or has not survived. A large mill was built at Kōloa in either 1840 or 1841. The two views of Kōloa plantation reproduced here were done in 1841 and attributed to Dr. Robert W. Wood who came to Honolulu in 1839 to head the Hospital for American Seamen; and as a capitalist began investing in Koloa Plantation in 1840. However, a similar illustration on page 191 of Ethel Damon's book Koamalu is attributed to Dr. Wood's wife, Delia, in 1850.

A number of stone barracks for native workers was described in 1846 by Chester S. Lyman, and these can be seen in the upper right corner of Metcalf's survey. Dr. Wood, in describing Koloa Plantation in a letter to Minister of Foreign Affairs Robert Crichton Wyllie in 1849, notes that there are 32 stone dwellings and 30 thatched dwellings, including stone apartment buildings, or barracks, for 10 and 9 Hawaiian families respectively. According to Arthur C. Alexander, the author of a definitive study of the plantation, these structures represent the model of plantation housing that was adopted throughout Hawaiian sugar plantations.[5] Other buildings adjacent to the barracks are the manager's house, a plantation store, and two structures of unknown use. The bulk of the cane lands was off this map toward the upper left.

The *ahupua'a* of West Kōloa was designated a *konohiki* land in the *mahele* and given to Moses Kekūāiwa, for Mataio Kekūanao'a, and is referenced by Land Commission Award 7714-B and Royal Patent 6714. It is described as consisting of 8,620 acres, but would have been reduced somewhat by the eighty *kuleana* awards within Kōloa. Moses Kekūāiwa died in 1848, and his lands went to his father, Mataio Kekūanao'a.

In the early years of the Kōloa Plantation, much of the cane was purchased from growers who leased smaller parcels of land. Dr. Robert Wood held a lease on the property shown here with his name in 1847. Reuben Tinker, a missionary, grew cane on a twenty-five acre farm across from the mill, for which he ultimately conveyed his lease to Ladd and Company. His property is identified with the Hawaiian spelling of his name, R.Tineka. Two other parcels that are shown belonged to Paulo Kānoa, the governor of Kaua'i beginning in 1847. Penciled annotations in Kānoa's lands show the name of James Young, whose full name was James Young Kānehoa, one of the original members of the Land Commission.

In December 1847, the lease to Ladd & Co., Reuben Tinker, and John Stetson, then comprising Koloa Plantation, was not renewed. Instead, a fifty year lease on 2,200 acres was issued to Dr. Robert Wood on September 25, 1848. On Metcalf's survey, Wood's name appears on one parcel of much smaller size. Dr. Wood operated Koloa Plantation successfully until 1872. In 1853 Dr. Wood bought the 3,263 acre "ahupua'a" of Pa'a, awarded to Jonah Pi'ikoi in the "mahele" and 690 acrees of the "ahupua'a" of Weliweli which adjoined Koloa. According to Alexander, Dr. Wood uncharacteristically passed on the chance to buy the "ahupua'a" of Koloa outright from Lot Kamehameha in 1863 who instead sold it to Robert Chrichton Wyllie.

A number of other properties are also indicated on Metcalf's map. One large parcel is defined as "aina o ka pake," or land belonging to a Chinese. This is probably the 28.9 acres of land sold in 1848 to a man identified as Aleon under Grant 92 in the *Index of All Grants and Patents Land Sales*. Samuela Namee received a grant of 6.83 acres in 1851 as Grant 526. These two transactions reflect the fact that existing leases were recognized during the *mahele*, and that lessees in some cases were given the opportunity to purchase the rented land.

The American Board of Commissioners for Foreign Missions received 825.35 acres in Kōloa under Land Commission Award 387 and Royal Patents 1936 and 1600. This property is identified on Metcalf's survey as "*aina o ka misionari o* Kōloa."

There is no firm date that can be assigned to Metcalf's survey. Based on somewhat contradictory internal evidence—information about transaction dates relating to the properties shown on the map—it seems that it may have been made in 1847.

68 Kōloa Mill Boiling House

Painting by Robert Wood; 15.8 x 29.2 cm.
Source: Grove Farm Homestead Museum

69 Kōloa Mill Residences

Painting by Robert Wood; 16 x 24.2 cm.
Source: Grove Farm Homestead Museum

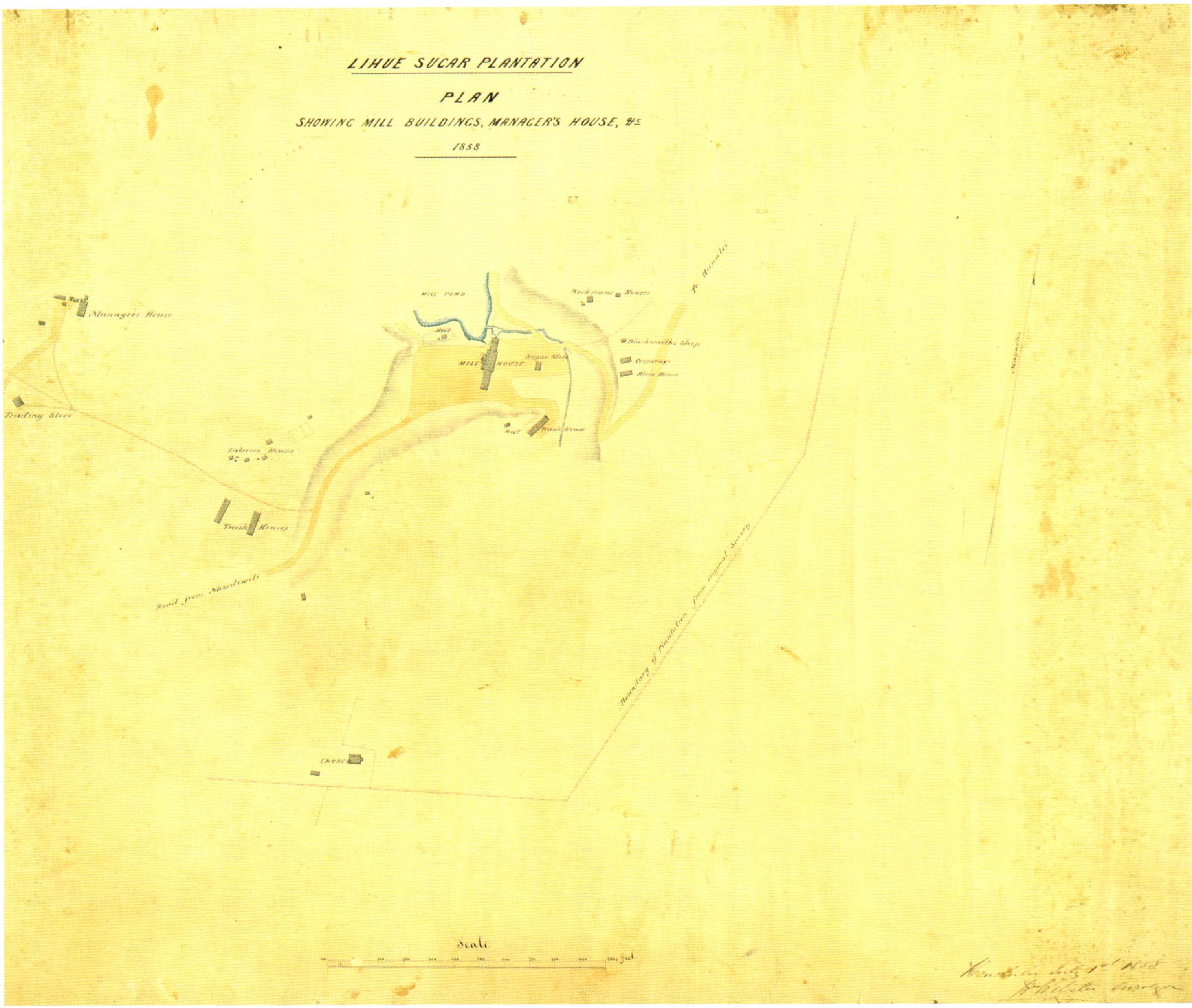

70 Līhu'e Sugar Plantation, Kaua'i

Surveyed by William Webster, July 1, 1858.
54 x 75 cm.; scale 1:2,400
Source: Hawaii State Survey Division.

Līhu'e Sugar Plantation

Līhu'e, the county seat of the island of Kaua'i, is one of the plantation towns that arose out of the sugar industry. The well-watered and relatively flat coastal plain of southeastern Kaua'i was the home of three major plantations: Kōloa, Grove Farm, and Līhu'e. This part of the island developed as the major focus of economic activity on the island, partly due to the presence of two natural harbors, Nāwiliwili and Hanamā'ulu, that were developed to serve the sugar industry. The area depicted in this survey is in the heart of the present town of Līhu'e.

71 Līhu'e Mill, ca. 1865
Photographer unknown.
Source: Grove Farm Homestead Museum

Webster's plan illustrates Līhu'e plantation nine years after it was established in 1849 by Henry Augustus Pierce, Judge William Little Lee, the chairman of the Land Commission, and Charles Reed Bishop. It was probably the best-capitalized and most-modern plantation at that time in all Hawai'i. A steam-powered mill was built in 1853 at Līhu'e Plantation, the first use of steam power on a Hawaiian sugar plantation. Another important innovation at Līhu'e was created in 1856, when William H. Rice completed the ten-mile-long Hanamā'ulu Ditch, the first large-scale irrigation project for any of the sugar plantations.

Līhu'e plantation is in the *ahupua'a* of Nāwiliwili, which was given to Victoria Kamāmalu in the *mahele*. Kamāmalu received all of the *ahupua'a* in the *moku* of Puna from Hanamā'ulu to Kīpū.

Webster's survey of Līhu'e Plantation is notable for its details of the physical structure of an early plantation. In addition to portraying the mill itself, he shows the location of several "huts," the sugar store, and trash houses. The self-sufficiency necessary for a sugar plantation is reflected in Webster's portrayal of a blacksmith's shop, a cooperage, and a storehouse. The residences of the manager, laborers, and workmen are shown as segregated in three distinct areas. A few unidentified buildings are scattered in various places. A church is portrayed near the southern property line, clearly within the plantation itself.

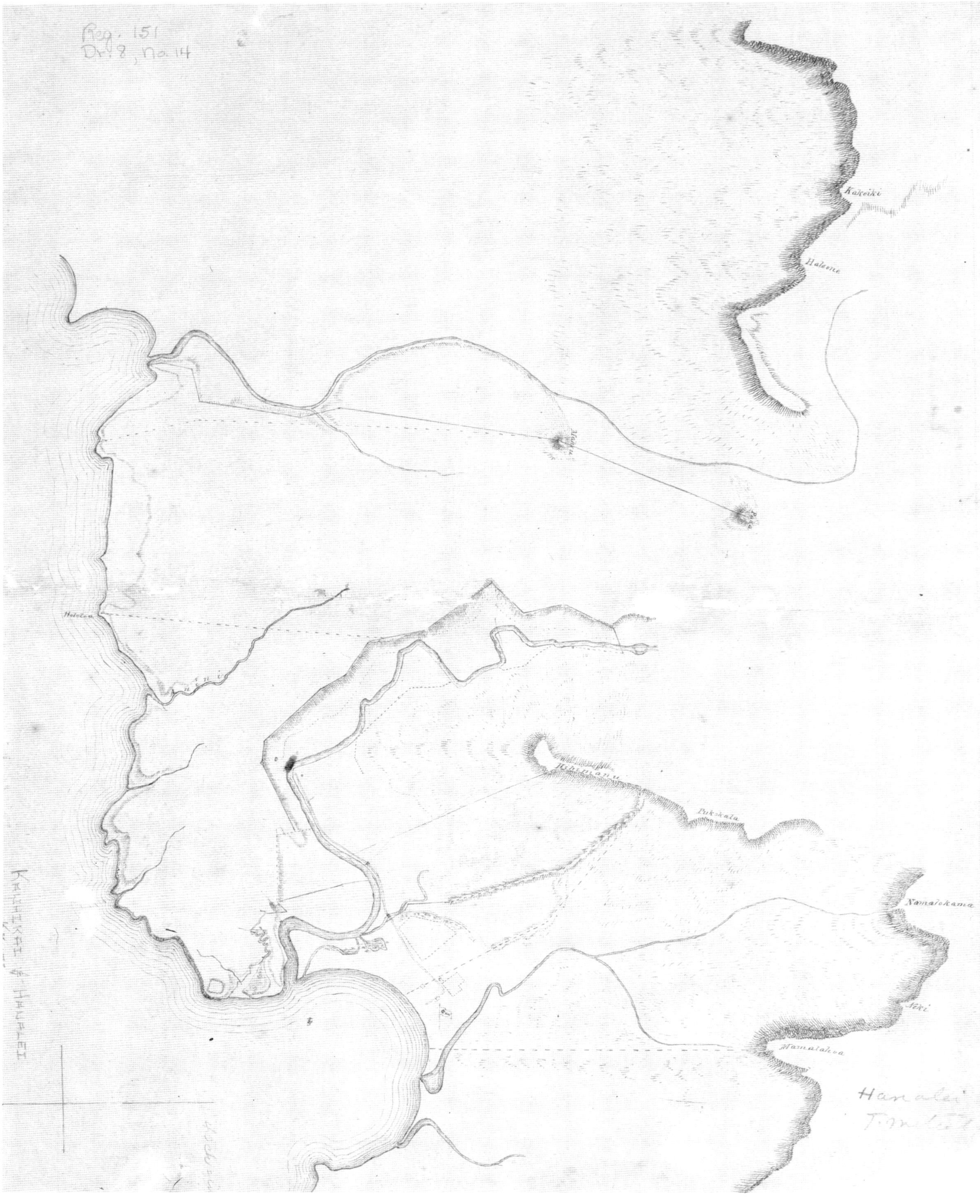

72 Kalihikai, Hanalei, and Wai'oli
Surveyed by Theophilus Metcalf, 1848

73 Hawaiian House, Mauka of Kalihiwai, ca. 1890

Source: Bernice Pauahi Bishop Museum, Honolulu, Hawai'i; negative number CP 54598.

Kalihikai, Hanalei, and Wai'oli

In this survey of the well-known area around Hanalei Bay on the north side of Kaua'i are the three *ahupua'a* of Kalihikai, Hanalei, and Wai'oli. In the *mahele*, Kalihikai was designated a *konohiki* land belonging to A. Keli'iahonui, the son of the hereditary *ali'i'ai moku* of Kaua'i, Kaumuali'i. Interestingly, this was the only *ahupua'a* on Kaua'i that was given to an *ali'i* native to the island, the rest being given to *ali'i* born and raised on the other islands. Hanalei was retained by Kauikeaouli as a crown land, and Wai'oli was assigned to the category of government land.

The geography of this portion of Kaua'i presented challenges to the surveyor in the *mahele* era. Kalihikai is the simplest of the three, consisting mostly of reasonably level land lying between two hundred and five hundred feet in elevation. It is about a mile and a quarter wide at the ocean and terminates at a point near the peak of Kapaka approximately four miles inland at an elevation of about one thousand feet.

Hanalei is a large *ahupua'a*, extending along the Hanalei River approximately eleven miles to the summit of Wai'ale'ale, the highest point on the island of Kaua'i. Much of the *ahupua'a* is comprised of mountainous terrain, including the gorge through which the Hanalei River descends toward the ocean. The mouth of the valley, however, is a broad alluvial plain that today is virtually covered with taro fields. There is also an extensive coastal plain lying between roughly one hundred and four hundred feet elevation.

Wai'oli stretches only about one-half mile along the shore of Hanalei Bay, but inland reaches a width of nearly two miles as it encompasses the drainage basins of three streams that form a large amphitheater-type feature. Approximately five miles inland, Wai'oli culminates near the southernmost tip of Māmalahoa Mountain at an elevation of 4,421 feet.

The rugged nature of the landscape in this area presented tremendous difficulties for surveyors of the *mahele* era. Consider, for instance, the boundary between Hanalei and Wai'oli. Starting at the shore of Hanalei Bay, it runs for about a half-mile across the flat plain, then begins tracing the ridge line between the drainage basins of the Hanalei River and the Wai'oli stream. From the peak of Pu'u Ki, the boundary rises along a sharp ridge nearly seven hundred feet in a horizontal distance of only one-half mile. At another point along this knife-edged ridge, the boundary rises nearly two thousand feet in a horizontal distance of just four thousand feet! Although the last mile or so is relatively flat, it defines the edge of a sheer, one-thousand-foot cliff along the eastern side of Māmalahoa Mountain.

Even in this day of helicopters, reaching these areas is not simple—and in Metcalf's era hauling equipment to these points would have been extremely difficult and dangerous. Consequently, Metcalf restricted his delineation of the boundaries of these *ahupua'a*. He did not show, for instance, the plateau atop Māmalahoa Mountain where Hanalei and Wai'oli join.

According to the *Indices of Awards*, there were nineteen *kuleana* grants within Kalihikai, fifty-five within Hanalei, and seventy in Wai'oli. It is interesting that Wai'oli, much smaller and with far less level land than Hanalei, had a significantly larger number of *kuleana* properties. Judging by the extensive taro farming that occurs in Hanalei today, it would seem that the well-watered valley must have been home to a considerable population in the original Hawaiian society. Archeological and other evidence, however, indicates that much of the low-lying land in Hanalei was not used extensively for agriculture in earlier times. The variety of taro grown by Hawaiians is reported to do poorly unless grown in cool water, and the broad plains of Hanalei did not have the slope necessary to keep water flowing continuously through the *lo'i*, or taro fields.

NIIHAU
Scale 2 inches to the mile
Scale of miles
Scale of chains

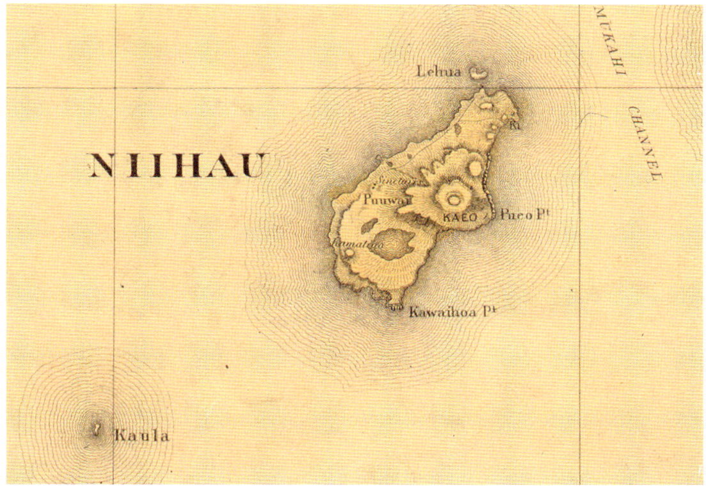

74 Ni'ihau, 1876

From: Map of the Hawaiian Islands by Hawaiian Government Survey, 1876. 98 x 154 cm.; scale 1:360,000.

Source: Library of Congress Geography and Map Division.

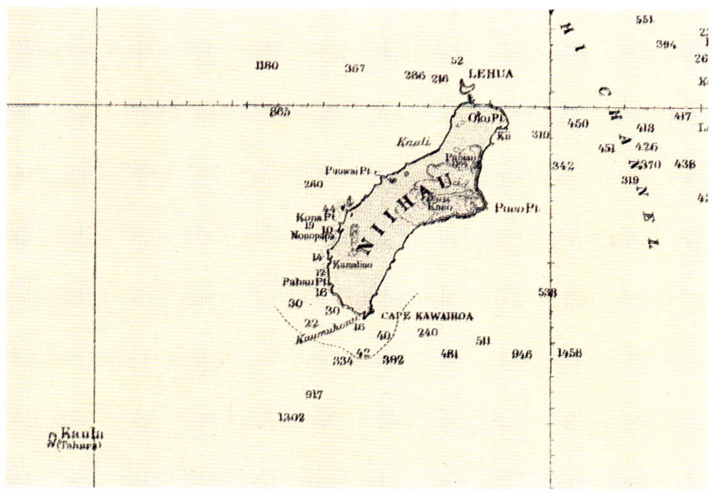

75 Ni'ihau, 1907

From: U.S. Coast and Geodetic Survey Chart 4100, 1907.
67 x 104 cm.; scale 1:600,000.
Source: Library of Congress Geography and Map Division.

76 Ni'ihau, 1865

Surveyed by George Norton Wilcox; tracing by M. D. Monsarrat, ca. 1875.
130 x 99 cm.; scale 1:31,680.
Source: Hawaii State Survey Division; reg. map no. 442.

Ni'ihau by George Norton Wilcox, 1866

Ni'ihau, the westernmost of the eight main islands of Hawai'i, has one of the simplest yet most controversial land histories of any of the islands. On February 23, 1864, by Royal Patent 2944, it was sold in its entirety by Kamehameha V to James M. and Francis Sinclair. Even before the sale, there was general dissatisfaction among the Hawaiian people with the amount of land that had been sold by Kauikeaouli and his two successors, Alexander Liholiho and Lot Kamehameha. In this particular transaction, nearly 47,000 acres—almost twice the amount granted to all *maka'āinana* as *kuleana* awards—was placed in the hands of non-Hawaiians. Just one month before the sale, residents of Ni'ihau petitioned the minister of Interior, G. M. Robertson, to be allowed to purchase the land instead of having it go to foreigners. The sale of Ni'ihau is occasionally cited as one of the reasons for the passage of an act by the legislature on January 3, 1865, that essentially forbade the king from alienating—selling—any additional lands, although other factors actually had more bearing on the creation of that act.

George Norton Wilcox is representative of many surveyors of this period who engaged in the occupation intermittently. Born in Hilo in 1839 and raised in Hanalei on Kaua'i, where his parents ran the Wai'oli mission, Wilcox studied at the Sheffield Scientific School at Yale University. Surveying was a regular part of the curriculum at the Sheffield School at that time, so it is not surprising that Wilcox worked at that trade intermittently for a few years upon his return to Kaua'i in 1863. In 1865, he accepted an assignment to go to Ni'ihau for three weeks to survey the recently sold land. For this, Wilcox was paid six dollars a day, a healthy sum that helped defray the purchase of Grove Farm Plantation, which he had acquired that same year.

Wilcox actually made the second survey of Ni'ihau, the first being done by a surveyor named J. Densison, who had been commissioned in 1833 by the missionary establishment to make a map of the islands. Denison's map did not survive, but copies of three portions, as drawn by Ursula Emerson, a missionary at Waialua on O'ahu (EMH, Ill. 61), have survived, including the map of Ni'ihau.[6] The most salient feature of Wilcox's survey is an enlarged representation of the southern part of the island. This area of Ni'ihau was thus overestimated in reference works for some years hence. The shape and extent of Ni'ihau was corrected in 1904 by W.D. Alexander. Because Wilcox's survey was made to document a legal transaction, it became part of the government records, unlike the earlier Emerson map, which was apparently known only to a few people. Consequently, it served as the basis for the portrayal of Ni'ihau on the 1876 "Centennial" map of the kingdom, the first official map of the islands published by the Hawaiian Government Survey.

In absolute measure, Wilcox's error of approximately 14,000 acres in the size of Ni'ihau is rather egregious, but it should be remembered that he was a young man with little experience at undertaking a project of this magnitude. To a large extent, such young men engaged in surveying work not as their primary occupation but because there was plenty of work, few surveyors, and good pay.

76 George Norton Wilcox, 1865
Photographer unknown.*80*.
Source: Grove Farm Homestead Museum.

6 Comments on the Mahele

UNCERTAINTY is a recurring theme in this volume. Much remains to be deciphered or rediscovered about how Hawaiians used their land before contact was made with Western society. And much research is needed before we will fully understand the extent to which land practices were altered in the period between the arrival of Europeans and the adoption of private ownership of land. The fact that the different legislative, political, and administrative acts that comprise the *mahele* occurred within a period of extreme social upheaval further complicates the effort to make definitive statements about many aspects of this momentous era in Hawaiian history.

Even the surveys that are the subject of this book are somewhat cryptic in what they tell us about the process of transforming Hawaiian society to a system based on private land ownership. But it is important to note that studying the *mahele* from the perspective of cartographic history does offer some insights that seem to have been overlooked by other historical treatments.

Probably the most definitive conclusion that can be drawn from a review of *mahele* surveying practices is that the advocates and implementors of the *mahele* did not put in place all the governmental institutions necessary to support the concept of private land ownership. The blame for this rests squarely on the foreigners who were pushing the Hawaiian government to permit the private ownership of land. This includes the missionary advisors, the foreign emissaries, the business community, and the men who accepted major governmental positions. Since the Hawaiians had no direct experience with the concept, they were dependent on the foreigners who explained the system to them.

Many scholars have clearly shown that the net result of the *mahele* was a significant loss of economic independence for the Hawaiian people. In addition to owning the mercantile businesses that had evolved, people of non-Hawaiian descent came to control much of the agricultural sector of the economy. And the loss of political independence in the 1890s was related to the changes in the economic structure of the islands.

In light of the disastrous impact of the *mahele* on the long-term economic and political well-being of the Hawaiian people, a perplexing question arises: Why weren't better preparations made before embarking on something as ambitious the *mahele*? From what has been learned in compiling this volume, the actions of the foreigners cannot be blamed on sinister motives. Instead, three basic factors largely account for the insufficient groundwork: an unquestioning faith in the superiority of private land ownership; a lack of appreciation for the complexity of private land ownership as a multi-faceted social concept; and the inexperience of key individuals upon whom Kauikeaouli and other Hawaiians had to rely for advice. It can be said with confidence that little thought was given to the need for or the availability of qualified surveyors.

The General Nature of the Surveys of the Mahele

One would like to say that the surveys of the *mahele* provide us with a thorough record of land use patterns during that era. Unfortunately, the surveying work that was done left an often incoherent picture of this profoundly important period in Hawaiian history. By their nature, surveys are sparse documents. The graphic images are usually quite simple, often no more than a few straight lines, and the accompanying textual descriptions follow conventions that did not encourage revealing descriptions of the inhabitants or lands.

Nonetheless, the surveys of the *mahele* are important, for when used in conjunction with other source material they substantially broaden our understanding of pre-*mahele* land use patterns and also shed more light on the *mahele* itself and its outcome. Making any effective use of the surveys from this era must be done with some knowledge of three fundamental areas: how surveys are made; the cultural foundations of the concept of private land; and the abilities and limitations of the men who actually measured and drew the plats that became records of *mahele* awards.

Surveying as an Integral Part of Land Ownership

How would things have been different if the need for surveying had been addressed before embarking on the *mahele*? Early American history offers an answer, and it should have been well-known at least to the Americans in Hawai'i. The march of American society across the midwestern states in the early nineteenth century was channeled through a systematic, well-defined framework of range and township units, in which major surveying landmarks were established by the government before land was sold to the public. Although it was unoccupied lands being surveyed and sold in the United States, the situation was analogous to conditions in Hawai'i in that no documentation of previous ownership or occupancy existed.

It would not have been necessary to create a rigid framework of ranges and townships in order to prepare for the *mahele*. However, it would have been possible to establish benchmarks to which surveyors could have referenced their plats. In areas where a heavy volume of land transactions was expected, such as around Honolulu and Lahaina, establishment and mapping of benchmarks and baselines prior to the actual transfer of land would have greatly aided record keeping. Given the limited financial and manpower resources that were available in Hawai'i, however, it is hard to imagine very much being done to establish a surveying infrastructure before the *mahele*. Surveying deficiencies had the greatest impact on lands then less valuable but later incorporated into the plantations that developed in the islands. It was these properties that were most likely lost after the objects that marked their boundaries were plowed under to make way for the fields of sugar cane, rice, and pineapple.

Although it is hard to imagine how any preparation for surveying could have been carried out in Hawai'i, it is clear that some of the problems that followed the *mahele* might have been less severe if there had been a substantial body of competent surveyors to help out.

Discord and Surveyors in the Mahele

The *mahele* might have proceeded differently if more attention had been paid to determining the exact nature of the taxation system that would be needed afterward. It must be remembered that in pre-*mahele* Hawai'i, there was little distinction between government and the economy. Taxation and economic activity were essentially one and the same. The people who were in control of the society—its government in essence—were supported directly through the control of land. The large class from which *ali'i* and *konohiki* came oversaw economic production and also was responsible for the smooth functioning of all cultural activities. Roles of people were well understood, and everyone knew from where their sustenance would come.

The disastrous effect of the *mahele* on the Hawaiian people can be understood if looked at from the perspective of the *ali'i*. Although many of them received vast acreage, the *maka'āinana* obtained their own land and were freed of the obligation to work for the king and the *ali'i*. How, then, were the *ali'i* to support themselves? And how were they to support the retinues they had always maintained?

And what about the civil service? Where was the money to come from to pay the ministers, judges, clerks, and laborers necessary to run a government? In pre-*mahele* Hawai'i, the people were obligated to provide the food that supported the government workers. Abolition of the *poalima* tax eliminated the basic foundation of government. Yet nothing was done to substitute for this source of income. The sale and leasing of land in a country as small as Hawai'i could not possibly sustain the government for very long. Usable land is a scarce resource in Hawai'i. Taxes on land were a staple revenue in the Western countries from which the concept of private land ownership was introduced to Hawai'i. Yet no coherent tax code seems to have been suggested by the people who were advocating the adoption of private land ownership.

It is unclear to what extent this issue was formally discussed by those in power, but it appears that many Hawaiians were keenly aware of the potential problems. The lack of a clear vision of the nature of taxation and the economy in the post-*mahele* society left the Hawaiian people in a quandary that had a direct impact on the surveyors of the era. The letters of Curtis Jere Lyons and Henry Lyman clearly indicate that there were numerous conflicts between *ali'i*, the *konohiki*, and *maka'āinana* when it came time to survey the claims for *kuleana* lands. It is impossible to say how pervasive this discord was because only a few surveyors seem to have reported it. But, as noted earlier, many of the surveyors may not have been cognizant enough of the Hawaiian language and culture to be fully aware of the concerns of the people for whom they were working.

Conflicts between the classes were inevitable. To the *ali'i*, giving the *maka'āinana* land was tantamount to making them economic equals. For the *maka'āinana*, fighting to gain the largest possible *kuleana* would be a way to achieve some security in an era of much uncertainty. Literature about the *mahele* is dotted with vague references to disputes between the various classes, and there are specific references to instances where *ali'i* and *konohiki* discouraged *maka'āinana* from filing their claims. Overall, however, there is a scarcity of documentation about such discord that makes it unwise to suggest the extent to which the classes were at odds with each other. But from Lyons and Lyman we have direct evidence that discord did make itself known to the surveyor in the field.

Where Did All the Land Go?

One fact is virtually uncontested: Little of the land of Hawai'i is still in the possession of those people who are largely of Hawaiian ancestry. There is no question that the *mahele* did more than provide for the transition to private ownership of land; it resulted in non-Hawaiians quickly gaining possession of large tracts of land and ultimately acquiring much of the most valuable land, other than that retained by the government. The *mahele* even proved a failure in that last regard, because the crown and government lands were lost to Hawaiians after the overthrow of the monarchy.

Others have detailed the story of how the land was ultimately lost. In this volume, however, it has been pointed out that a principal mechanism for providing land to the *maka'āinana* has not been given sufficient consideration—the sale of land to individual Hawaiians. This process was considered from the start to be an integral part of the *mahele,* and a review of the official records clearly shows that in a large number of transactions much land was sold to people with Hawaiian names. As noted earlier, the *maka'āinana* did not receive much acreage as a result of *kuleana* awards, but they apparently did acquire considerable acreage through purchase.

How much land did Hawaiians acquire through purchase? This is an area which awaits some valuable research. It would be interesting to know more than just the number of transactions and the acreage involved. What was the geographical distribution of these sales—by island, *moku*, and *ahupua'a*? Did these sales represent quality agricultural land? What happened to the parcels over time? To what extent were they subsequently sold to non-Hawaiians and how quickly did this occur? Just as important, who were these Hawaiians who purchased land? Were they *maka'ainana*, or *ali'i* and *konohiki*? Because so much land was involved in these sales, such questions must be answered before conclusions are drawn about the fairness of the *mahele* to the Hawaiian people.

Youth, Experience, and Motivation

Many of the European and American proponents of private land ownership in Hawai'i were young and relatively inexperienced. A naivete regarding the complexity of land ownership, or at least a lack of appreciation for the sophisticated nature of the Hawaiian land tenure system, is reflected in some of the ideas such people suggested. For example, William Little Lee published the suggestion that Hawaiians trade their scattered properties among themselves so that each could come forward with a claim for one contiguous piece of property. But the Hawaiian tradition of disconnected property made great sense in that it provided individuals with access to a variety of farmland suitable for different types of crops. If the mainstay of life was to be agriculture, as it always had been, then combining fields did not make as much sense to Hawaiians as it did to Lee. At the time of the *mahele*, there was no real sense that a new economy was to be built on plantation agriculture, so the suggestion for single-unit farms had little practical application.

A major factor in understanding the outcome of the *mahele* involves the motivations of the various parties involved. Kauikeaouli certainly struggled with many issues as he grew into his reign. Coming of age as a nation's monarch would not be easy for any individual, but to do so within a society undergoing the massive transformations occurring in Hawai'i was an almost overwhelming burden to bear. Many motivating factors, not just one, no doubt influenced his decision to embark on the *mahele.*

As for the *ali'i* and *konohiki*, their motivations spanned a much greater range of emotions and concerns than can be addressed here. With their society changing drastically and rapidly, their concerns probably would have been changing with events. Among the upper level of *ali'i*, who received most of the land in the *mahele*, the politics of power was certainly too complex to summarize neatly. Indeed, that is largely the focus of Lilikalā Kame'eleihiwa's important work on the *mahele*, *Native Lands and Foreign Desires*.

For others involved in the *mahele*, motivations could have run a wide gamut. Missionaries who acted as surveyors and land agents seem to have done so partly because they felt such duties were consistent with their chosen work, and partly because they felt the need to give their families financial security or benefits that could not otherwise be obtained. The fact that descendants of missionary families ended up controlling a significant portion of the land in Hawai'i is not in itself an indication that the missionaries themselves had planned to wrest control of the country from the people they came to help.

It is impossible to say what motivated the variety of foreigners who came to Hawai'i and quickly found themselves in positions of power and trust. Many of the men who were instrumental in the implementation of the *mahele*, such as William Little Lee, William Richardson, and John Ricord, did not realize significant personal gains from the *mahele*. Did they act out of altruistic beliefs, or were they merely responding to the best of their abilities in positions that they found themselves in through fate or chance?

Two of the more interesting people involved in the surveying of the *mahele* were the youngsters Curtis Jere Lyons and Henry Munson Lyman. For their motivation, it is probably better to look at them as young men excited by the adventure and responsibility that fell their way. Their upbringing in missionary families seems to have endowed them with a sense of fairness and concern regarding the *maka'āinana*, but it is probably not necessary to invoke that part of their past to explain their diligence in working at their accidental professions. They were simply intelligent young men who were given a chance to do some work that gave them money and independence, common enough aspirations for young people.

Given these differing motivations, it is probably futile to seek underlying themes.

The Need for the Hawaiian Government Survey

As an exercise in surveying, it can be argued that the *mahele* did not go well. Records of property were not of uniform quality, and there was no central authority within the government that was charged with or capable of ensuring that all surveys agreed with each other.

From the perspective of government, there were other problems directly related to surveying and mapping. As the government became increasingly dependent on the revenue from sales and rental of government lands, the lack of maps made land management very difficult. It was impossible to rent large tracts of land if the *kuleana* awards within them were not clearly identified. Deciding the value of land for taxation, sale, or lease was difficult if acreage had not been determined. If existing leases were not recorded on accurate maps, there was the danger of leasing parcels to more than one party. And without maps, it was hard to determine what good lands were available for sale or lease.

The only way to clear up this massive disorder was to create a government agency charged with mapping the lands of Hawai'i and providing oversight of all public and private surveying within the islands, and in 1870, the Hawaiian Government Survey was established to do just that. The next volume in this series will look at how the men of the Hawaiian Government Survey went about their work and will examine the maps and charts that constitute its legacy.

Notes

Chapter 1: The Hawaiian Use of Land

1. Kuykendall, p. 9.
2. Culliney, p. 324; Kirch p. 292.
3. Coulter, p. 3.
4. Kuykendall, pp. 9–10.
5. Pukui, p. 18.
6. Kirch and Sahlins, v. 1, p.
7. Pukui, p. 251.
8. Dole, pp. 4–5.
9. Kamakau, p. 7.
10. Dole, 1892, p. 5 .
11. Kamakau, p. 6.
12. Lyons, 1858.
13. Lyons, 1903, p. 29.
14. Kamakau, p. 6.
15. King, p. 19.
16. Lyons, 1903, p. 29.
17. Pukui, p. 59.
18. King, unpaged glossary.

Chapter 2: Surveying and the Private Ownership of Land

1. Byron, p. 155.
2. Kuykendall, p. 159.
3. Kuykendall, p. 122.
4. Kirch and Sahlins, v. 1, p. 133.
5. Kirch and Sahlins, v. 1, p. 133.
6. Kotzebue, v. 1, p. 319.
7. Kotzebue, v. 1, p. 340.
8. Kotzebue, v. 1, p. 341.
9. Arago, 1823, v. 2, p. 119.
10. McCormac, p. 394.
11. Wall, p. 73.

Chapter 3: The Mechanics of the Mahele

1. Chinen, p. 7.
2. Kuykendall, v. 1, p. 161.
3. Chinen, p. 8; Kuykendall, v. 1, p. 269.
4. Chinen, p. 9.
5. Kuykendall, v. 1, p. 282.
6. Chinen, p. 16.
7. Chinen, p. 25.
8. Chinen, p. 26.
9. Wyllie to Judd, Nov. 19, 1849
10. Morgan, p. 133.
11. Kame'elehiwa, p. 295.
12. Alexander, Arthur Chambers, 1920, p. 24.

Chapter 4: The Surveyors of Hawai'i

1. Alexander, Mary, p. 284.
2. Emerson, p. 169.
3. Kirch and Sahlins, v. 1, p. 168.
4. Emerson, p. 142.
5. Emerson, 142.
6. Emerson, 142.
7. Lyons, 1890.
8. Lyons, 1890.
9. Lyons, 1890.
10. Lyons, 1890.
11. Lyons, 1890.
12. Lyons, 1890.
13. Lyman, p. 205.
14. Lyman, p. 232.
15. Lyons, 1890.
16. Lyons, 1851.
17. Lyons, 185-.
18. Lyons, 1851.
19. Lyons, 1851.

Chapter 5: Surveys from the Mahele

1. Kuykendall, p. 283.
2. Hawaii. Foreign register, Vol. 2, p. 112.
3. Hawaii. Foreign Testimony.
4. Wylie, p. 69.
5. Alexander, Arthur, *Koloa*, p. 32.
6. Forbes, p. 58.

Bibliography

Adler, Jacob and Robert M. Kamins.

1986 *The Fantastic Life of Walter Murray Gibson: Hawaii's Minister of Everything*. Honolulu: University of Hawaii Press.

Akana-Gooch, Collette I.

1983 *John S. Emerson: A Missionary's Desire for Land*. masters thesis, University of Hawaii.

Alexander, Arthur Chambers.

1937 *Koloa Plantation 1835–1935*. Honolulu: Star-Bulletin.

1920 "Land Titles and Surveys in Hawaii." In *Hawaiian Planters Record*. August, 1920. pp. 67–78.

Alexander, Mary Charlotte.

1934 *William Patterson Alexander*. New Haven: Yale University Press.

Alexander, William DeWitt.

1882 *A Brief History of Land Titles in the Hawaiian Kingdom*. Honolulu: P. C. Advertiser Co.

1898 "The Maker of the Hawaiian Flag." *Sixth Annual Report of the Hawaiian Historical Society*. Honolulu. pp. 8–11.

1908 "Early Improvements in Honolulu Harbor." *Fifteenth Annual Report of the Hawaiian Historical Society*. Honolulu. pp. 13–35.

Arago, Jacques.

1823 *Narrative of a Voyage Around the World in the Uranie and Physicienne Corvettes, Commanded by Captain Freycinet, During the Years 1817, 1818, 1819, and 1820....* .London: Treuttel and Wurtz.

Beechert, Edward D.

1991 *Honolulu: Crossroads of the Pacific*. Columbia: University of South Carolina Press.

Bushnell, O. A.

1993 *The Gifts of Civilization: Germs and Genocide in Hawaii*. Honolulu: University of Hawaii Press.

Chinen, Jon Jitzuso.

1958 *The Great Mahele: Hawaii's Land Division of 1848*. Honolulu: University of Hawaii Press.

1961 *Original Land Titles in Hawaii*. Honolulu: Jon Jitzuso Chinen.

Coulter, John Wesley.

1931 *Population and Utilization of Land and Sea in Hawaii*. 1853. Honolulu: Bemice P. Bishop Museum. (Bulletin 88.)

1935 *A Gazetteer of the Territory of Hawaii*. Honolulu: University of Hawaii. (Research Publications No. 11.)

Creighton, Thomas H.

1978 *The Lands of Hawaii: Their Use and Misuse*. Honolulu, University Press of Hawaii.

Culliney, John L.

1988 *Islands in a Far Sea: Nature and Man in Hawaii*. San Francisco: Sierra Club Books.

Daws, Alan Gavan.

1966 *Honolulu—The First Century: Influences in the Development of the Town to 1876*. Ph.D. thesis, University of Hawaii.

1968 *Shoal of Time*. Honolulu: University Press of Hawaii.

Day, Arthur Grove.

1984 *History Makers of Hawaii: A Biographical Dictionary*. Honolulu, Mutual Publishing.

Devaney, Dennis M., et. al.

1982 *Kaneohe: A History of Change*. Honolulu: Bess Press.

Dole, Sanford B.

1892 *Evolution of Hawaiian Land Tenures*. Honolulu: Hawaiian Historical Society . (Hawaiian Historical Society Papers No. 3.)

Dorton, Lilikala.

1986 *Land and the Promise of Capitalism: a Dilemma for the Hawaiian Chiefs of the 1845 Mahele*. Ph.D. dissertation, University of Hawaii.

Dutton, Meirie K.

1953 *William L. Lee*. Honolulu: Loomis House Press.

Earle, Timothy K.

1978 *Economic and Social Organization of a Complex Chiefdom: the Halelea District. Kauai. Hawaii*. Ann Arbor: Museum of Anthropology, University of Michigan.

Emerson, Oliver Pomeroy.

1928 *Pioneer Days in Hawaii*. New York: Doubleday, Doran & Co.

Emory, Kenneth P.

1924 *The Island of Lanai*. Honolulu: Bishop Museum (Bulletin 12).

Forbes, David W.

1992 *Treasures of Hawaiian History*. Honolulu: Hawaiian Historical Society.

Hackler, Rhoda E. A.
1989 *R.W. Meyer Sugar Mill. Moloka'i: Its History and Restoration.* Moloka'i Museum and Cultural Center.
Hawaii. Foreign Register.
Hawaii. Foreign Testimony.
Hawaii. Office of the Commissioner of Public Lands.
1916 *Index to All Grants and Patents: Land Sales.* Honolulu, Paradise of the Pacific Print.
1929 *Indices of Awards Made by the Board of Commissioners to Quite Land Titles in the Hawaiian Islands.* Honolulu: Star Bulletin Press.
Hobbs, Jean.
1935 *Hawaii: a Pageant of the Soil.* Palo Alto: Stanford University Press.

Jarves, James Jackson.
1846 *Investigation by Command of the King and Premier of the Hawaiian Islands.* Honolulu: James Jackson Jarves.
Joesting, Edward.
1984 *Kauai: The Separate Kingdom.* Honolulu: University of Hawaii Press
Judd, Bernice.
1936 "Koloa: A Sketch of Its Development." *Forty-Fourth Annual Report of the Hawaiian Historical Society for the Year 1935.* Honolulu: Hawaiian Historical Society.

Kamakau, Samuel Manaiakalani.
1976 *The Works of the People of Old; Na Hana a ka Po'e Kahiko.* Honolulu: Bishop Museum Press.
Kame'eleihiwa, Lilikala.
1992 *Native Land and Foreign Desires.* Honolulu: Bishop Museum Press.
Kelly, Marion.
1956 *Changes in Land Tenure in Hawaii 1778–1850.* masters thesis, University of Hawaii.
King, Robert D.
1942 "Hawaiian Land Titles." *Real Property Appraisers Manual.* Honolulu.
Kirch, Patrick Vinton.
1986 *Feathers Gods and Fishhooks.* Honolulu: University of Hawaii Press.
Kirch, Patrick V. and Marshall Sahlins.
1992 *Anahulu: The Anthropology of History in the Kingdom of Hawaii.* 2 vols. Chicago: University of Chicago Press.
Kotzebue, Otto von.
A Voyage of Discovery, into the South Sea and Beerings Straits, for the Purpose of Exploring a North-east Passage Undertaken in the Years 1815–1818.... London: Longman, Hurst, Rees, Orme, and Brown.
Krauss, Bob with William P. Alexander.
1983 *Grove Farm Plantation.* 2nd ed. Palo Alto: Pacific Books.
Kuykendall, Ralph S.
1953–1957 *The Hawaiian Kingdom.* 3 vols. Honolulu: University of Hawaii Press.

Lam, Maivan Clech.
1989 "The Kuleana Act Revisited: The Survival of Traditional Hawaiian Commoner Rights in Land." *Washington Law Review.* April 1989. pp 233–288, v. 64.
Linnekin, Jocelyn.
1983 "The Hui Lands of Keanae: Hawaiian Land Tenure and the Great Mahele," *Journal of the Polynesian Society.* Vol. 92, No. 2, June 1983. pp.169–188.
1987 "Statistical Analysis of the Great Mahele," *Journal of Pacific History.* Vol. 22, No. 1, January, 1987. p. 15–33.
Lydgate, John Mortimer.
1914 "The Vanishing Kuleana." *Hawaiian Almanac and Annual for 1915.* Honolulu: Thos. G. Thrum." p. 103–109.
Lyman, Henry Munson.
1906 *Hawaiian Yesterdays.* Chicago: A.C. McClurg Co.

Lyons, Curtis Jere.
1851 Letter, February 1851, from Lyons to George Robertson.
1851 Letter, March 12, 1851, from Lyons to George Roberston.
1858 Letter, October 4, 1858.
1890 "Forty Years Ago". Typescript in Hawaiian Mission Childrens Society Library dated 1889 or 1890.
1902 "Land Matters in Hawaii" *Report of the Territorial Surveyor.*
1903 *A History of the Hawaiian Government Survey with Notes on Land Matters in Hawaii.* Honolulu: Hawaiian Gazette.
1926 "Early Surveying and Land Allotting in Hawaii" *Honolulu Advertiser.* February 24, 1926, p. 13.

Mann, Cline.
1978 "Hawaiian Lands—A Brief Description." typescript.
"Manoa Valley: Descriptive, Historic, Legendary."
1891 *Hawaiian Almanac and Annual for 1892.* Honolulu: Press Publishing, pp. 110-116.
McCormac, Jack C.
1991 *Surveying Fundamentals, 2nd ed.* Englewood Cliffs, NJ: Prenctice Hall.
"Metes and Bounds."
1939 *The Sales Builder.* Vol 12, No. 7, July 1939, pp. 2–14.
Meyer, Charles S.
1982 *Meyer and Molokai.* Alden, Iowa: Graphic-Agric Business.
Missionary Album.
1969 Honolulu, Hawaiian Mission Children's Society.
Morgan, Theodore.
1965 *Hawaii. A Century of Economic Change 1778–1876.* Cambridge: Harvard University Press.

Pukui, Mary Kawena and Samuel H. Elbert.
1986 *Hawaiian Dictionary; Hawaiian-English, English-Hawaiian; Revised and Enlarged Edition.* Honolulu: University of Hawaii Press.

Robertson, George N.
1856 "Final Report of the Board of Commissioners to Quiet Land Titles." *Annual Report of the Minister of the Interior.* Honolulu: Government Printer.

Schmitt, Robert C.
1973 *Missionary Censuses of Hawaii.* Honolulu: Bishop Museum. (Pacific Anthropological Records 20).
Spaulding, Thomas Marshal.
1923 *The Crown Lands of Hawaii.* Honolulu. (University of Hawaii Occasional Paper No. 1).

Tabrah, Ruth.
1976 *Lanai.* Honolulu: Island Heritage.
1987 *Niihau: The Last Hawaiian Island.* Kailua: Press Pacifica.

Wall, Walter E.
1906 "Survey Department;" *Annual Report of the Governor of Hawaii for 1906.* Washington, GPO.
White, David H.
1971 *Children of the Land: A Historical Analysis of Land Alienation Among the Hawaiian People Until 1898.* senior honors thesis, University of Hawaii.
Wise, John H.
1933 "The History of Land Ownership in Hawaii." *Ancient Hawaiian Civilization.* Honolulu: Kamehameha Schoo!s, pp 77–89.
Wyllie, Robert Crichton.
1848 *Answers to Questions Proposed by His Excellency. R.C. Wyllie.* Honolulu: Government Printer.
1849 Letter to Gerritt P. Judd, November 19.

Index

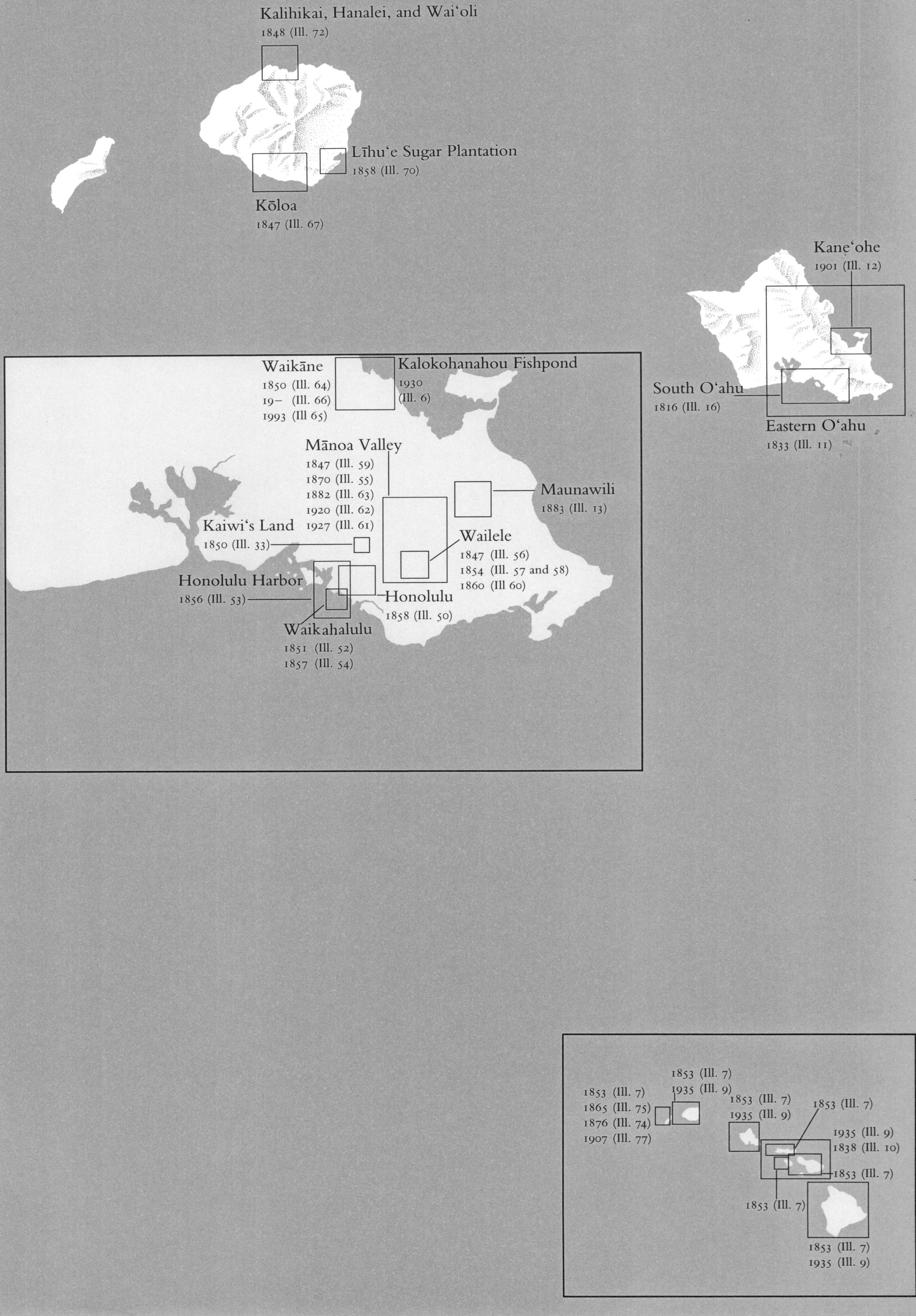

Kalihikai, Hanalei, and Wai'oli
1848 (Ill. 72)
Līhu'e Sugar Plantation
1858 (Ill. 70)
Kōloa
1847 (Ill. 67)
Kane'ohe
1901 (Ill. 12)
South O'ahu
1816 (Ill. 16)
Eastern O'ahu
1833 (Ill. 11)
Waikāne
1850 (Ill. 64)
19– (Ill. 66)
1993 (Ill 65)
Kalokohanahou Fishpond
1930
(Ill. 6)
Mānoa Valley
1847 (Ill. 59)
1870 (Ill. 55)
1882 (Ill. 63)
1920 (Ill. 62)
1927 (Ill. 61)
Maunawili
1883 (Ill. 13)
Kaiwi's Land
1850 (Ill. 33)
Wailele
1847 (Ill. 56)
1854 (Ill. 57 and 58)
1860 (Ill 60)
Honolulu Harbor
1856 (Ill. 53)
Honolulu
1858 (Ill. 50)
Waikahalulu
1851 (Ill. 52)
1857 (Ill. 54)
1853 (Ill. 7)
1865 (Ill. 75)
1876 (Ill. 74)
1907 (Ill. 77)
1853 (Ill. 7)
1935 (Ill. 9)
1853 (Ill. 7)
1935 (Ill. 9)
1853 (Ill. 7)
1935 (Ill. 9)
1838 (Ill. 10)
1853 (Ill. 7)
1853 (Ill. 7)
1853 (Ill. 7)
1935 (Ill. 9)